International Federation of Clinical Chemistry
Recommendations and Related Documents, Volume 1

International Federation of
Clinical Chemistry

Recommendations and Related Documents
Volume I, 1978 – 1983

Quantities and Units
Quality Control
Enzymes
Proteins
Instrumentation
Diagnostic Kits and Reagents
pH and Blood Gases
Education

Editor Nils-Erik Saris

W DE G

Walter de Gruyter · Berlin · New York 1984

Professor Dr. Nils-Erik Saris
Department of Medical Chemistry
University of Helsinki
Siltavuorenpenger 10
SF 00170 Helsinki 17, Finland

CIP-Kurztitelaufnahme der Deutschen Bibliothek

International Federation of Clinical Chemistry:
Recommendations and related documents / International Federation of Clinical Chemistry / ed. Nils-Erik Saris. – Berlin ; New York : de Gruyter
NE: Saris, Nils-Erik [Hrsg.]

Vol. 1. 1978–1983 : quantities and units, quality control, enzymes, proteins, instrumentation, diagnostic kits and reagents, ph and blood gases, education. – 1984.
ISBN 3-11-008766-9

Library of Congress Cataloging in Publication Data

International Federation of Clinical Chemistry.
Bibliography: v. 1, p.
Contents: v. 1. 1978–1983––
1. Chemistry, Clinical––Quality control––
Collected works. 2. Chemistry, Clinical––Standards––
Collected works. I. Saris, Nils-Erik. II. International Federation of Clinical Chemistry. [DNLM: 1. Chemistry, Clinical. 2. Quality Control. QY 90 I612r]
RB40.I62 1984 616.07'5 84-16994
ISBN 3-11-008766-9 (v. 1)

Typesetting and Printing: Mercedes-Druck, Berlin. – Binding: Lüderitz & Bauer, Buchgewerbe GmbH, Berlin. – Printed in Germany

Preface

The first document from the International Federation of Clinical Chemistry (IFCC) was a monograph – R. Dybkær and K. Jørgensen: Quantities and Units in Clinical Chemistry, Including Recommendation 1966 of the Commission on *Clinical Chemistry of the International Union of Pure and Applied Chemistry and of the International Federation for Clinical Chemistry* (Munksgaard, Copenhagen, and Williams & Wilkins, Baltimore, MD, 1967). In 1975 *Clinica Chimica Acta* agreed to publish IFCC Sections, and *Journal of Clinical Chemistry and Clinical Biochemistry* also published these from the beginning. Since then a great number of IFCC documents and recommendations have been published, all of them in these two journals, and in some other leading journals of the field as well, e.g. *Clinical Chemistry, Clinical Biochemistry, Journal of Clinical Laboratory Automation* and *Clinical Chemistry Newsletter.* Translations into Italian, Spanish and Russian have also appeared in national or regional journals.

In 1983 a milestone was passed when the Expert Panel on Nomenclature and Principles of Quality Control in Clinical Chemistry completed its task and the Council approved the six recommendations it had produced. The great interest in its work is evidenced by the fact that National Groups in Italy, Spain and Latin America have produced booklets containing translations of these documents even before some of them had been finalized.

The present book contains the set of approved recommendations on quality control in clinical chemistry. It also contains the approved recommendations on quantities and units in clinical chemistry which have had a substantial impact on the field. The Executive Board of the IFCC decided that other completed projects also should be included. It should be emphasized, however, that an approved recommendation or a completed project may not be the final word on the matter. As progress continues, some texts may become at least partially obsolete, and additional aspects may require attention. Nevertheless, it is hoped that the publication of this volume will aid clinical chemists and related professionals by bringing together the relevant documents scattered throughout the literature as they were published in the cooperating journals over the years.

The book contains references to all documents and recommendations published in English. In addition it contains information on the activities and terms of reference of the various Expert Panels. It is hoped that this will enhance the usefulness of the collection. Future volumes will appear at suitable intervals as projects are completed.

Nils-Erik Saris

Contents

Introduction

The International Federation of Clinical Chemistry

René Dybkær, President 1979–1984

We have to live today by what truth
we can get today and be ready
tomorrow to call it falsehood.
William James

With this first collection of published papers from the International Federation of Clinical Chemistry (IFCC), it may be useful to present the structure and functions of the Federation because they influence the product.

There is no official *definition* of "Clinical Chemistry", but the following may serve the purpose: The paraclinical discipline called clinical chemistry is a special branch of medicine dealing with measurement and interpretation of the physico-chemical condition and dynamics in healthy and diseased humans, thus contributing to a pathophysiological understanding and thereby to the prophylaxis, diagnosis, therapy, prognostication and research of disease. Such designations as "clinical biochemistry" and "chemical pathology" cover the same field and, in practice, there are substantial overlaps between clinical chemistry and other clinical laboratory sciences, especially haematology, immunology, physiology and pharmacology.

It is characteristic that clinical chemists are asked to measure many different types of quantities of widely different nature, in large numbers, within a short time, and on small specimens presenting great matrix problems. Therefore, the clinical chemical laboratory, perhaps more than other analytical laboratories, has had to utilize a wide selection of quite different techniques evolved from physics, chemistry, radiochemistry, biochemistry, immunology and microbiology. Due to the potential seriousness of "false" results, great stress has been put on quality control, and to cope with the increasing workload sophisticated mechanization and electronic data processing have been applied.

The analytical part of clinical chemistry, in spite of its many exciting challenges, must not obscure our main goal: to participate in the elucidation of the physiological and disease process of each patient. In fact, the development of clinical chemistry should be guided by the effort to solve pathobiochemical problems.

With the increasing complexity of our discipline, the need to exchange and pool knowledge and experience resulted in the *formation* of National Societies, starting with the Societé Française de Biologie Clinique in 1942. Soon, a formalized relation between the national groups became natural, and an international embryo was formed in 1952 as another aspect of the clinical chemical activites within the International Union of Pure and Applied Chemistry (IUPAC). In 1967 the IFCC achieved its independence, yet maintaining strong and useful links with the IUPAC.

The *goals of IFCC* are "to advance the science and practice of clinical chemistry and to enhance its service to health and medicine" in the interest of all peoples of the world. The *objectives* are to encourage high professional standards and the formation of National Societies as well as to promote international cooperation and free exchange of professional information among clinical chemists and with neighbouring disciplines.

The Federation comprises a governing *Council* composed of one National Representative for each member society, of which there are at present 43. The Council elects the officers of the *Executive Board*, which conducts business in the interims between the triennial Council meetings at the International Congresses of Clinical Chemistry.

The IFCC is convinced that many industrial or professional corporate groups share its objectives, and since 1978 such organizations can become Corporate Affiliates of the Federation. The more than 30 groups in this category form an important part of the IFCC expertise and potential.

In view of the many tasks of the Federation and the common interests with other professional organizations, our international relations have been continuously expanded to minimize duplication and to utilize resources optimally. Our closest contact is with the IUPAC and its Clinical Chemistry Division, with which we enjoy a pleasant and fruitful collaboration as evidenced by the joint papers in this volume. An increasingly important and demanding collaboration has evolved due to our official relation to WHO, whose worldwide channels of communication and stature increase the impact of clinical chemical joint projects on the national health services, including those in countries that do not yet have IFCC Members.

The Federation also values its excellent working relations with the International Committee for Standardization in Haematology, the International Organization for Standardization, the International Union of Biochemistry, the International Union of Immunological Societies, and the World Association of Societies of Pathology, as well as its participation in a general information exchange between many standards-setting bodies such as the European Committee for Clinical Laboratory Standards, the International Organization of Legal Metrology, the National Bureau of Standards (US), and the National Committee for Clinical Laboratory Standards (US).

In addition to participating in the international work of the Federation, some clinical chemical societies have formed regional groups, and we are happy to acknowledge collaboration with our colleagues in the Latin American Confederation of Clinical Biochemistry and the Asian and Pacific Federation of Clinical Biochemistry.

In the structure of IFCC, four bodies are concerned with organizational matters: the *Awards Committee*, the *Congress Committee* with its Regional Committees for Europe and for African, Mediterranean and Near Eastern countries, the *Nominations Committee*, and the *Publications Committee*. Three groups supervize and conduct studies that may lead to IFCC Recommendations or Documents: the *Education Committee*, the *Office of Reference Materials and Methods* (ORMM) and the *Scientific Committee* (SC). These bodies may form Expert Panels (EPs) or Working Groups (WGs) on specific subjects. At present, the ORMM has a WG on Reference Methodology using Mass Spectrometry and the SC has an Analytical Section comprising EPs on Apolipoproteins, Core Clinical Laboratory, Enzymes, Evaluation of Diagnostic Reagent Sets, Instrumentation, pH and Blood Gases, Quality Control, Quantities and Units as well as a WG on Plasma Protein Standardization; the Clinical Section includes EPs on Drug Effects, Nutrition, Theory of Reference Values, and a WG on Effectiveness of Quantities. In addition, there are two groups to address specific problems put to IFCC by WHO: a Task Force on Essential Tests for WHO Intermediate Laboratories and another on Simple Tests.

The many colleagues working within this structure and its numerous national and international relations produce the means to achieve our objectives in the form of congresses and other international and regional meetings, national or regional courses in theoretical and practical aspects of our field, and study reports of various kinds. It is the last mode of operation that is our concern here.

The present volume is the direct outcome of the interaction between the draft-producing Committees, Expert Panels, Working Groups and Task Forces, and an extensive net of commentators from all over the world. The Federation stresses the usefulness of *consensus* documents and, therefore, follows a rigorous process of draft circulation to Associate Members of Expert Panels, both from National Societies and Corporate Affiliates, as well as to the National Representatives and Corporate Affiliate Representatives, to pertinent international bodies and individual experts. After revision, the paper may be published as an IFCC Document or as a provisional recommendation for further comment, revision, and Council ballot to be republished as an IFCC-Recommendation. All such papers appear in *Clinica Chimica Acta* and *Journal of Clinical Chemistry and Clinical Biochemistry;* selected papers are also published in other journals.

If the reader finds the present documents widely differing in character, this is to be expected when an international organization tries to produce a wide spectrum of offerings so as to provide useful material for nearly everyone.

The Nature of IFCC Documents and Recommendations

Robert Zender, Chairman of the Scientific Committee 1979–1982

IFCC documents are intended either to become recommendations approved by the Council, or to summarize present-day international expertise in a given field. In the latter case approval of the Council is not always solicited when the document is non-controversial (a list of references for instance) or when a reasonable consensus was reached by draft circulation of proposals that will very likely be modfied later (an analytical method for instance).

Recommendations and documents are prepared with two aims in view: to promote quality in clinical chemistry practice, and to introduce norms (written standards) attuning methods, securing control materials and harmonizing modes of expression. IFCC documents can be categorized into those pertaining to clinical practice and the interpretation of laboratory data, those more closely related to biochemistry and physiology, and those touching on methodological aspects, i.e. analytical chemistry.

IFCC documents can also be classified according to their intended use, education for instance, and they likewise fall into categories according to their intended readership. The latter classification is not obvious, and some documents were misunderstood by readers who found them esoteric, obvious or simply irrelevant to their particular field. We would like to lay stress on this aspect of the problem.

Some recommendations are intended more for standardization groups or scientific committees of national societes than for each and every member. For instance reference methods proposed by the Expert Panel on Enzymes are designed to serve as international "yardsticks" rather than as routine bench methods. According to prevailing conditions and usages it is the responsibility of the local, national or regional organizations to implement such reference methodology or to harmonize existing technology with the yardstick. Fortunately, our world is varied and diverse; thus yardsticks are necessary, and only an international body such as the IFCC can provide them with some chance of succeeding.

Other documents and recommendations are written with a view to filling in the existing gap between our discipline and the more basic sciences. Papers, for instance on spectroscopy, or on physicochemical quantities, prepared by the Expert Panel on Quantities and Units or the Expert Panel on pH and Blood Gases, are of this kind. These documents may seem esoteric to most of us, but is it not our personal responsibility to expand and further our knowledge? Furthermore such documents prepared by clinical chemists, both MDs and PhDs, are also addressed to our colleagues in the basic disciplines, who have found some of them quite useful because, as Tom Whitehead wrote some years ago, clinical chemistry is a view from a bridge in which both banks, medicine and chemistry, seem closer and more familiar. Such documents serve also to improve the image of clinical chemistry.

The Expert Panels of IFCC

Most of the work in the preparation of documents and recommendations is carried out in the Expert Panels of the Scientific Committee, the Education Committee, and the Office of Reference Materials and Methods. While the expertise in the Panels is truly international, specific tasks may be carried out by local Working Groups. The Expert Panels which have successfully produced published documents or are in the process of doing so, are presented briefly below. The general field of activity is indicated by the name of each Expert Panel. The terms of reference define the tasks in greater detail. The Panels are presented in the order in which they were established.

Expert Panel on Quantities and Units

The Panel was created in 1968 under the chairmanship of R. Dybkær (Copenhagen, DNK). The present Chairman is H. P. Lehman (New Orleans, USA). Its terms of reference are:

1. Examination of recommendations on nomenclature and rules of quantities, kinds of quantities and units from pertinent international bodies.
2. Active cooperation with relevant international bodies concerned with making recommendations in the same field, especially the International Union of Pure and Applied Chemistry, Section of Clinical Chemistry, Commission of Quantities and Units in Clinical Chemistry.
3. Drafting recommendations or advocating use of the recommendations of other bodies relevant to the panel's general field of activity. The final approval of a recommendation will rest with the Scientific Committee and ultimately with the IFCC Council.
4. As far as possible the proposals of the Panel should be in accordance with the decisions of international bodies allied to clinical chemistry.
5. Suggesting ways and means of implementing the international use of the approved recommendations.

Expert Panel on Proteins

The Panel was created in 1969 under the chairmanship of T. Freeman (London, GBR). In 1980 the Panel was dissolved and its tasks transferred to a Working Group on Protein Standardization under the chairmanship of B. G. Johansson (Malmö, SWE). According to the terms of reference, the following point should be examined for each protein:

1. Definition of protein, nomenclature and units.
2. Definition of a reference preparation: physical and chemical properties (including reactive properties), purity, storage conditions, stability, production and control.
3. Description of one (possibly two) reference method(s) for the determination of the protein in pertinent biological materials.
4. Availability and distribution of reference preparation.
5. As far as possible, the proposals of the Panel should be in accordance with the decisions of international bodies allied to clinical chemistry.
6. Implementation of the international use of recommendations (when finally approved).

Expert Panel on Nomenclature and Principles of Quality Control

The Panel was created in 1969 under the chairmanship of J. Büttner (Hannover, DEU). It was dissolved in 1983 after having completed its tasks. Its terms of reference were:

1. Preparation of a list of English terms and definitions in the field of quality control in clinical chemistry (QC). Translation into German and French.
2. Intralaboratory QC techniques. Description and use. Collection of data and statistical treatment. Self-prepared and commercial reference materials.
3. Interlaboratory QC techniques. Descriptions and use. Collection of data and statistical treatment.
4. Preparation of a general plan for standardization of clinical chemical methods utilizing results from items 2 and 3.
5. Preparation of a list of recommended reference materials and corresponding methods.
6. Recommendations on intra- and interlaboratory QC and on minimal requirements of routine accuracy and precision.

7. Recommendation to manufacturers for preparation of commercial reference materials.
8. As far as possible the proposals of the Panel should accord with decisions of international bodies allied to clinical chemistry.
9. Implementation of the international use of recommendations (when finally approved).

Expert Panel on the Theory of Reference values

The Panel was created in 1970 under the chairmanship of R. Gräsbeck (Helsinki, FIN). The present Chairman is H. E. Solberg (Oslo, NOR). Its terms of reference are:

1. Nomenclature. Kinds of reference values ("normal values").
2. Ways of selecting the individuals constituting a group yielding data for study.
3. Factors influencing reference values.
4. Transformation of data. Calculation of parameters characterizing the distribution of data.
5. Combination of data-yielding groups.
6. Ways of presenting reference values for research workers and for clinicians.
7. Ways of presenting single measurements as related to appropriate reference values.
8. As far as possible the proposals should be in accordance with the decisions of relevant international bodies.
9. Advice on the implementation of the international use of recommendations (when finally approved).

Expert Panel on Enzymes

The Panel was created in 1971 under the chairmanship of G. N. Bowers (Hartford, Conn., USA). Its present Chairman is R. Rej (New York, N.Y., USA). Its terms of reference are:

1. Definition of selected enzyme(s) nomenclature and units.
2. Definition of properties of a reference preparation for the enzyme(s), production and control included.
3. Description of one (possibly two) reference method(s) for the determination of the enzyme(s) in pertinent biological materials.
4. As far as possible, the proposals of the Panel should be in accordance with the decision of international bodies allied to clinical chemistry.
5. Suggestions for implementation of the international use of recommendations (when finally approved).

Expert Panel on Instrumentation

The Panel was created in 1974 under the chairmanship of F. L. Mitchell (Harrow, GBR). Its present Chairman is T. D. Geary (Adelaide, AUS). Its terms of reference are:

1. The following points should be examined for the major classes of instruments (such as centrifuges, pipettes, chromatographs, mass spectrometers, multichannel automatic analyzers) and recommendations made.
2. Definition of terms, with special reference to performance characteristics, for:
   a) instruments in general
   b) the main classes of instruments.
   These definitions should be in accord with the definitions and decisions of other Expert Panels and should take note of the decisions of other bodies allied to clinical chemistry.
3. Prepare one or more protocols for the testing and evaluation of instruments, with particular reference to the ease of use, performance, safety and susceptibility to interference from external sources such as temperature, humidity, dust, vibration, electricity supply, etc.
4. Recommend standards of safety for instruments used in clinical chemistry laboratories, with due emphasis upon chemical, electrical, mechanical, microbiological and radioactive hazards.
5. Recommend standards for the presentation and content of instruction manuals, with special emphasis on:
   a) preventive maintenance
   b) repair facilities, particularly for laboratories remote from manufacturers workshops in developing countries.
6. Make general recommendations upon the requirements for instruments used in developing countries.
7. Advice upon the implementation of these recommendations.

Expert Panel on pH and Blood Gases

The Panel was created in 1977 under the chairmanship of H. Maas (Utrecht, NLD). Its terms of reference are:

1. Review of definitions of quantities used for description of the components related to hydrogen ions, carbon dioxide and oxygen in systems of clinical chemical importance.
2. Recommendations concerning preferred quantities and their definitions.
3. Recommendations of preferred units.
4. Recommendations of symbols for quantities and units.

5. Establishment of reference values.
6. Recommendations concerning terminology for values outside the reference interval.
7. Recommendations concerning reference materials, reference methods and quality control.
8. Recommendations concerning sample methodology and instrumentation.
9. Recommendations concerning nomograms and algorithms.
10. The proposals of the Panel should be in accordance with decisions of international bodies allied to clinical chemistry and there should be agreement with other Expert Panels in areas of common interest as far as recommendations are concerned.

Expert Panel on Immunoassay

The Panel was created in 1977 under the chairmanship of J. Landon (London, GBR). It has been temporarily merged with the Expert Panel on Quantities and Units in Clinical Chemistry. Its terms of reference are:

1. The Panel is concerned with all aspects of immunoassays and of related analytical techniques employing natural circulating proteins or other proteins as the binding reactant (the reactant may be labelled with an isotope, an enzyme, a phage, a fluorescent or chemiluminescent molecule, etc. or it may be unlabelled).
2. Working out of terminology
3. Definition of properties of reference materials. Recommendation for their preparation. Recommendation for standardization and quality control.
4. Recommendation for the obtainment of reference values. Compilation of established reference values.
5. Recommendation for the preparation of reagents. Description of supplies. Protocol for the assessment of kits and reagents.
6. Recommendation for the requirement and design of instruments and of data handling facilities. Protocol for their evaluation.
7. Recommendation for education and training. Contents of programmes.
8. Close collaboration with other Expert Panels with the World Health Organization and with other international bodies such as international scientific societies. As far as possible the proposals of the Expert Panel should be in accordance with the decisions of relevant IFCC or other international bodies.
9. Advice on the implementation of the international use of recommendations (when finally approved).

Expert Panel on Drug Effects in Clinical Chemistry

The Panel was created in 1978 under the chairmanship of G. Siest (Nancy, FRA). Its terms of reference are:

1. The Panel is mainly concerned with the understanding and monitoring of the effects of drugs *in vivo*, rather than with the study of their interferences with clinical chemistry methodologies.
2. Selection or development of appropriate terminology, nomenclature and classification of analytical and biochemical effects of drugs. Tentative list and classification of effecting drugs.
3. Protocols for the systematic investigations of drug effects on particular clinical chemical parameters, allowing a discrimination between *in vitro* and *in vivo* effects.
4. Critical review of *in vivo* effects of drugs, particularly on enzyme levels. Recommendations as to the choice of appropriate biochemical assays for monitoring the clinical responses to particular drugs.
5. Close cooperation with other Expert Panels. The Panel should suggest ways and means for the implementation of their recommendations. As far as possible, the proposals should be in accordance with the decisions of relevant IFCC or other international bodies.

Expert Panel on Evaluation of Diagnostic Reagent Sets

The Panel was created in 1979 under the chairmanship of J. Logan (Ottawa, CAN) to continue the work of an Expert Group on Diagnostic Kits and Reagents that was formed with M. Rubin (Washington, D.C., USA) acting as Chairman in 1978. Its terms of reference are:

1. The Panel is concerned with diagnostic reagent sets or kits, by which is meant two or more different clinical or general laboratory materials (excluding reconstitution materials), with or without other components, packaged together and designed for the performance of a procedure for which directions are supplied with the package. Clinical laboratory materials are substances which are intended by the manufacturer as an aid in the laboratory control of patient management and which are not administered to the patient.
2. Recommendations for specifications of labelling, package, etc.
3. Recommendations for evaluation, general guidelines.
4. Recommendations for selection and evaluation of performance of reagent sets, for individual parameters.
5. The terminology used should be in accordance, as far as possible, with existing IFCC documents, particu-

larly those produced by the Expert Panels on Quantities and Units, Quality Control and Theory of Reference Values in Clinical Chemistry.

6. Close cooperation should be maintained with other Expert Panels, particularly those concerned with overlapping fields such as the Expert Panel on Instrumentation and that on Immunoassay. A clear delineation of activities is encouraged.

7. The Panel may, with the agreement of the Scientific Committee, enter into and/or fulfil contracts with or otherwise cooperate with relevant international bodies such as the World Health Organization, so that the documents produced would be of mutual benefit to the World Health Organization and IFCC.

8. Advice on implementation of the international use of the recommendation.

The Education Committee

The Education Committee was created in 1975 under the chairmanship of Professor Martin Rubin (USA). Its present Chairman is Professor Morton K. Schwartz (USA). The Committee works closely with the IUPAC Commission on the Teaching of Clinical Chemistry. The goals of the Committee are to develop educational and training requirements and programs for all levels of professional and technical personnel in Clinical Chemistry in both developing and developed countries.

The terms of reference of the Education Committee are:

1. To gather information on the state of education in clinical chemistry in the world and to disseminate this information.
2. To make recommendations on the content and standards of education programmes internationally.
3. To stimulate, sponsor or undertake educational programmes especially in developing countries.
4. To stimulate and assist the organization of educational activities such as workshops, short courses, symposia and congresses in clinical chemistry.
5. To stimulate and assist in the development of national societies in clinical chemistry.
6. To provide consultants in clinical chemistry education.
7. To collaborate with other national and international bodies whose aims are helpful in the development of clinical chemistry education.
8. To develop methods for evaluation of educational programmes and graduates of such programmes.
9. To develop the accreditation of educational programmes and certification and licensure of practitioners of clinical chemistry.

The current and most recent projects of the Committee include development of a Master of Science in Clinical Chemistry Course outline, implementation of an IFCC Visiting Lectureship, preparation of a syllabus for training of technical personnel, publication of a Reading List and creation of an Audio-visual library. The committee is also in the early stages of forming cooperative projects to develop Educational Material based on activities of Expert Panels.

Approved Recommendations and Related Documents

I. Expert Panel * on Quantities and Units in Clinical Chemistry

and the

Union of Pure and Applied Chemistry
Section of Clinical Chemistry
Commission on Quantities and Units in Clinical Chemistry

Approved Recommendations (1978)

* Titular Members

B. H. Armbrecht, Beltsville (USA) 1967–1975
R. Dybkær, Copenhagen (DNK) 1967–1977
(Chairman 1967–1975)
R. Herrmann, Gießen (DEU) 1971–1979
K. Jørgensen, Copenhagen (DNK) 1967–1973
P. Métais, Strasbourg (FRA) 1967–1975
J. C. Rigg, Wageningen (NLD) 1973–1977
O. Siggaard-Andersen, Copenhagen (DNK) 1975–1979
R. Zender, La Chaux-de-Fonds, (CHE) 1975–1979
Chairman

Associate Members

B. H. Armbrecht, Beltsville (USA) 1975–1979
K. Jørgensen, Copenhagen (DNK) 1975–1979
P. Métais, Strasbourg (FRA) 1975–1979

Titular Members and Associate Members are terms used by the International Union of Pure and Applied Chemistry

Quantities and Units in Clinical Chemistry

Prepared for publication by R. Dybkær

Preface

The commission on Quantities and Units in Clinical Chemistry is a part of the Section on Clinical Chemistry of the International Union of Pure and Applied Chemistry (IUPAC). The Expert Panel on Quantities and Units[1]) is a part of the Committee on Standards of the International Federation of Clinical Chemistry (IFCC). These two bodies, the Commission and the Expert Panel, have worked on this document, the former mainly concerned with basic philosophy, the latter with problems of implementation.

The aim has been to bring clinical chemical nomenclature into line with evolving international scientific language, especially by drawing upon the recommendations of the International Committee of Weights and Measures, the International Union of Biochemistry, the Technical Committee 12 of the International Organization for Standardization, and, of course, IUPAC (see References).

The tentative version of the present publication appeared as an IUPAC–IFCC (yellow) *Information Bulletin*, Appendices on Tentative Nomenclature, Symbols, Units, and Standards, No. 20, February 1972. As a consequence of later decisions by international bodies concerned with nomenclature, a considerable number of comments, and new deliberations by the Commission and Expert Panel (see Appendix), a Recommendation 1973 contained many, mostly, small, modifications from the tentative version. The Recommendation 1973 was approved by IUPAC in 1973 and published in Pure and Applied Chemistry 1974, *37*, 517–546. The IFCC

1) Members as above

Council in 1975 did not accept the text on account of another view on quantities and units in enzymology presented in the IFCC Expert Panel on Enzymes' Provisional Recommendation (1974) (13). At a joint meeting in Strasbourg 1976 the Expert Panel on Quantities and Units, the Commission on Quantities and Units in Clinical Chemistry, and the Expert Panel on Enzymes reconciled their concepts, which were further modified according to the advice of the IUPAC Interdivisional Committee on Nomenclature and Symbols. The present document incorporates the resulting changes in names and definitions of kinds of quantities in enzymology.

Where it conflicts, the present document supersedes the larger IUPAC–IFCC Recommendation 1966, its translation into Spanish, and the IUPAC Recommendation 1973.

This Recommendation has been approved by IUPAC, and by the Council of IFCC in February 1978.

Contents

Abbreviations of references

CBN	Commission on Biochemical Nomenclature (IUPAC–IUB)
CCC	Commission of Clinical Chemistry (IUPAC; before 1967)
CGPM	Conférence Générale des Poids et Mesures
CQUCC	Commission on Quantities and Units in Clinical Chemistry (IUPAC; after 1967)
EPQU	Expert Panel on Quantities and Units (IFCC Committee on Standards)
ICSH	International Committee for Standardization in Hematology
IFCC	International Federation of Clinical Chemistry
ISO	International Organization for Standardization
IUB	International Union of Biochemistry
IUPAC	International Union of Pure and Applied Chemistry
QU-R66	Quantities and Units in Clinical Chemistry. Recommendation 1966 (IUPAC–IFCC)
QU-R73	Quantities and Units in Clinical Chemistry. Recommendation 1973 (IUPAC)
SCC	Section on Clinical Chemistry (IUPAC)
WAPS	World Association of (Anatomic and Clinical) Pathology Societies

1. Introduction

Recommendation 1966 of the Commission on Clinical Chemistry of the International Union of Pure and Applied Chemistry and of the International Federation for Clinical Chemistry (4) was published in 1967. This was the first effort of international organizations towards a rationally standardized presentation of chemical data from clinical laboratories.

At the IUPAC Conference in Prague in 1967, the newly created Section on Clinical Chemistry nominated a Commission on Quantities and Units in Clinical Chemistry (see footnote [4])). This body was joined in 1968 by the Expert Panel on Quantities and Units created by the Committee on Standards of the International Federation of Clinical Chemistry (see footnote [5])). One common task was the supervision of short versions of the Recommendation 1966 (4) = QU-R66. The present publication contains such a revised condensation in Sections 3, 4 and 5 which are now the official version in cases of conflict. Chemical background material is given in Section 2. Section 6 corresponds to the QU-R66 Part 5 with new kinds of quantities. Section 7 contains examples of clinical chemical quantities in a new generic form, which is further elaborated in a separate publication 'List of Quantities. Approved Recommendation 1978' (pp. 35–47).

The kinds of quantities treated in this document do not by far cover all aspects of clinical chemical work. It is hoped that additional recommendations will be prepared to alleviate this defect.

2. Clinical chemical language problems

2.1. The main task of the clinical chemist is the measurement of the chemical, biochemical, and sometimes physical *properties of patients.* These data are used by the clinician for diagnostic and therapeutic purposes. It is of paramount importance, therefore, that the *determinations requested* by the clinician are those *performed* by the laboratory and that each *answer* is understood correctly by the clinician.

In order to avoid misunderstandings a set of rules must be observed during transmission, i.e. the 'language' must be standardized.

2.2. Traditionally, there has been no accepted usage for the request and presentation of data, be it on the international, national or local level. This has been *inconvenient*, time-consuming and, in some instances, dangerous to the patient.

Another, very unfortunate result of the diverse – and often quaint – usages in clinical chemistry is a *barrier of incomprehensibility* between this discipline and the rest of science where chemistry, biochemistry, and physics have cooperated on a standardized language for some years.

2.3. The international clinical chemical bodies now try to eliminate the language part of the barrier by *adopting pertinent parts of the communication language constructed by related international science bodies.* The *Recommendation 1966* by CCC of IUPAC and IFCC is updated and summarized in Sections 3 and 4; to understand them, some chemical concepts are recapitulated.

2.4. The clinical chemical description of a patient requires initially the selection of **a system**, which may be arbitrarily chosen. *Examples:* the patient as such, his liver, blood, thrombocytes, hemoglobin.

2.5. **Transported matter** to or from the system may be regarded as a component or group of components entering or leaving the system or can be considered a separate chemical system. *Examples:* food absorbed, urine excreted, blood supplied to the liver.

2.6. In clinical chemistry, it is usually convenient or necessary to refrain from using the entire system for measurement; instead **a specimen** is obtained, i.e. a representative part of a system or, at least, a part obtained by a standardized procedure. *Examples:* specimens of 24-h urine, of venous blood, of serum from venous blood. The actual measurement is often on a representative **sub-specimen**.

2.7. A chemically definable substance in the form of atoms, molecules or ions is called **a chemical component of the system.** *Examples:* helium, water, and sodium ion respectively.

2.8. **The chemical name**, if practical and possible, should be a fully specified systematic name, preferably constructed according to IUPAC rules. *Examples:* carbamide rather than urea, sodium ion (not 'sodium').

2.9. **The formula for a molecule or an ion** is a symbolic representation indicating the atoms (and charges, if any) involved in the entity. *Examples:*

H_2O (sterically H\O/H with H above each bond), CH_3COOH, SO_4^{2-}

2.10. For complicated molecules or where a name alone may signify more species of particles than intended, it is necessary to supply **an elementary entity** (a formula unit or elementary unit), i.e. a chemical symbol indicating a relevant repetitive structure of the chemical component. *Examples:* N_2, N, Ca^{2+}, COOH (carboxyl), H_2O, C_6.

2.11. One goal of the clinical chemical description is to furnish information on the *patient's biochemical status*, be it normal or pathological. The chemical processes taking place in a living organism are governed by well-known laws, which are formulated in 'molecular' terms. Consequently, clinical chemical measurements of amounts of components or their concentrations should *preferably be expressed with reference to the groups of atoms, molecules or ions participating.*

2.12. An evident advantage for clinical chemistry of the *'molecular' concepts* (see 2.13) was realized rather early for some inorganic electrolytes: sodium ion, potassium ion, chloride, and hydrogen carbonate. Today, the concentrations of these components are usually given as 'molar concentrations' in millimoles per litre (or 'milliequivalents per litre'). For many organic and some inorganic substances, however, *mass concentration* is widely used, often with a volume of 100 millilitres (of solution) as denominator in the unit.

Mass concentration rarely yields biologically useful information; in fact, biological relationships are obscured and, consequently, not often contemplated. In contrast, *'molecular' concepts immediately expose functional relationships.* A few series of interdependent components[2] might suffice to illustrate this fact:

acetoacetate – acetone – β-hydroxybutyrate
chloride – bromide
adrenaline – noradrenaline – 4-hydroxy-3-methoxymandelate
calcium ion – phosphate
copper ion – ceruloplasmin
cholesterol – cholesterol ester
creatine – creatinine
bilirubin – bilirubin glucuronide
'base excess' – lactate – glucose
cystine – cysteine
'reduced' glutathione – 'oxidized' glutathione
glycerol – triglyceride
hemoglobin(Fe) – oxygen(O_2) – iron ion – transferrin
protein-bound iodine – triiodothyronine – thyroxine

2.13. **SCC and IFCC recommend that kinds of quantities of a 'molecular' nature should be used whenever an elementary entity can be defined for a component**, be it an electrolyte or a non-electrolyte, inorganic or organic, and whatever its size.

Consequently, of the kinds of quantities listed in Section 4, the following are preferable when describing 'amount' or 'concentration' of matter.

amount of substance (mole)
instead of mass (kilogram),

substance concentration[3] (mole per litre)
instead of mass concentration (kilogram per litre),

substance fraction (also called mole fraction) (unity)
instead of mass fraction (unity), volume fraction (unity), and molality (mole per kilogram).

This recommendation is supported jointly by the 1972 Recommendation from ICSH, IFCC and WAPS (6).

2.14. The *Recommendation 1966* states that **the litre is the preferred unit for volume** (see (3), Note 2) in expressing concentration in liquids and solids, where volume is the denominator. The present arbitrary use of microlitre, millilitre, decilitre (= 100 millilitres), and litre repeatedly results in recalculations and consequent errors.

3. Definitions of fundamental concepts of quantities

3.1. **Quantity**: a measurable, real property, physical or chemical, of a specified system. *Example:* substance concentration of glucose in the blood plasma of a stated patient.

3.2. **Kind of quantity**: the abstract concept of the property, common to a number of real phenomena (quantities). *Examples:* length, substance concentration, pressure.

Note: ISO, IUPAC, and IUPAP employ the term 'quantity' both for the abstract, 'non-addressed' concept of, e.g., substance concentration and for the 'addressed' measurable property given in 3.1.

3.3. **Unit**: a chosen reference quantity, which may be used for comparison of quantities of the same dimension. *Examples:* millimetre (e.g. for lengths), kilogram (e.g. for masses), mole (e.g. for amounts of substance), gram per litre (e.g. for mass concentration or mass densities).

Note: The magnitude of the quantity is not influenced by the choice of unit.

3.4. **Numerical value**: the number that gives the magnitude of the measured quantity when multiplied by the unit.

$$\text{Numerical value} = \text{quantity/unit} \qquad (1)$$

---

2) The names listed accord with current usage and are not those used in 'List of Quantities. Approved Recommendation 1978'.

3) See 4.11.6.

Note 1. The magnitude of a quantity should not be expressed as a numerical value alone, except perhaps for quantities having the dimension one, see 3.8. *Example:* Substance concentration of sodium ion in a specified serum is not '139' but '139 millimoles per litre'. The relative density of the same serum could be adequately expressed as 1,026, but the unit '1' can be stated: 1,026 × 1.

Note 2. The 'equal to' sign does not allow inference of the kind of quantity involved from inspection of the unit employed. *Example:* The unit, gram per litre, is common to measurements of mass concentration and mass density.

Note 3. The meaning of the numerical value is a statistical analytical problem beyond the scope of this document. In general, it is advisable to supply information about analytic variation and reference values ('normal values').

3.5. **Base kind of quantity**: a kind of quantity considered dimensionally independent of other base kinds of quantities, i.e. it is not defined by an equation containing other kinds of quantities. *Examples:* length, mass, amount of substance.

Note. Base quantities are chosen arbitrarily but according to a practical plan. The seven base kinds of quantities corresponding to the International System of Units (SI units) are shown in Table 1 (3.5) (12).

Tab. 1 (3.5). Base kinds of quantities and SI base units.

Name	Symbol	Dimension	Name	Symbol
Length	l	L	metre	m
Mass	m	M	kilogram	kg
Time	t	T	second	s
Electric current	I	I	ampere	A
Thermodynamic temperature	T	Θ	kelvin	K
Luminous intensity	I	J	candela	cd
Amount of substance	n	N	mole	mol

3.6. **Derived kind of quantity**: a kind of quantity characterized by an equation between base kinds of quantities. *Example:* mass concentration = mass/(length)^3.

3.7. **Dimension of quantity**. Any (kind of) quantity can be analysed into a product of dimensional factors, one for each of the dimensions, corresponding to the seven base kinds of quantities. *Example:* Mass concentration and density both have the dimension $L^{-3}M^1T^0I^0\Theta^0J^0N^0 = ML^{-3}$, they are equidimensional; the appropriate units mg/l, g/l, kg/m^3, etc. have the same dimension.

3.8. **Kind of quantity having the dimension one (1)**: *Examples:* Number (of entities), mass fraction, relative density all have the dimension $L^0M^0T^0I^0\Theta^0J^0N^0 = 1$. (Such kinds of quantities are often called 'dimensionless'.)

3.9. **Base unit**: a unit defined for a base kind of quantity. *Examples:* metre, kilogram, mole. See Table 1 (3.5).

3.10. **Derived coherent unit**: unit constructed from base units, exclusively. *Examples:* mole per kilogram, kilogram per cubic metre.

Note: Kinds of quantities having the dimension one use the coherent unit 'unity' (1).

3.11. **Derived non-coherent unit**: unit constructed from base units and numerical factors, which may be named and symbolized (see 3.12). *Examples:* milligram, mole per litre.

3.12. **Factors for units**: These factors which are part of the SI (12), create multiples of units and are often called by a name used as a prefix or a letter symbol before the unit symbol. *Examples:* megametre (Mm) for 10^6 metre, micrometre (μm) for 10^{-6} metre; see Table 2 (3.12).

Note 1. The twelve prefixes above the upper and below the lower broken lines in the table are especially recommended by CQUCC and EPQU because the systematic use of factors that are powers of ten with exponents that are simple multiples of three facilitates conversion procedures for results in different units.

Tab. 2 (3.12). SI prefixes denoting decimal factors.

Factor	Prefix	Symbol
10^{18}	exa	E
10^{15}	peta	P
10^{12}	tera	T
10^9	giga	G
10^6	mega	M
10^3	kilo	k
- - -	- - -	- - -
10^2	hecto	h
10^1	deca	da
10^{-1}	deci	d
10^{-2}	centi	c
- - -	- - -	- - -
10^{-3}	milli	m
10^{-6}	micro	μ
10^{-9}	nano	n
10^{-12}	pico	p
10^{-15}	femto	f
10^{-18}	atto	a

Note 2. These prefixes may not stand alone and must not be used together with the unit 'unity' (1), employed by quantities having the dimension one (cf. 3.10, Note). This is not practical and could be solved by designating a name and symbol to 'unity'.

4. Kinds of quantities and units

4.0. Only the most necessary information is given in connection with each *kind of quantity:* name, (type), symbol, and in some cases an identifying definition; the *unit* name, (type), symbol, and sometimes a definition. The abbreviations for type of kind of quantity and for unit are explained in Table 3 (4.0).

Tab. 3 (4.0). Codes used in sections 4 and 6

Code	Meaning
B	Base
D	Derived
DC	Derived coherent
DnC	Derived non-coherent
spec C	Special coherent

4.0.1. The italic symbols for a number of intensive kinds of quantities do not carry a subscript indicating the component; e.g. the former c_B for 'substance concentration of component B' is now given as c. In this context, probably, only mass density and mass concentration may be confused as both are symbolized ρ. When necessary, mass concentration is symbolized ρ_B or $\rho(B)$; other kinds of quantities may be specified analogously.

Recommended types of symbols for subunits and some corresponding forms that are not recommended [in brackets] are listed. Examples are given in Section 7, with codes and abbreviations explained in Tables 4 (7.3) and 5 (7.5).

4.1

Length (B)	*l*
Metre (B)	m
m	
mm	[cm][4])
μm	[μ, u]
nm	[mμ, mu]

Note 1. The spelling 'metre' is recommended in English (12). The American form is 'meter'.

Note 2. The order of magnitude changes by a factor of 10^{-3} between successive units listed.

4.2

Area (D)	*A*
Square metre (DC)	m^2
m^2	[cm^2][4])
mm^2	
μm^2	[μ^2][4])

Note. The order of magnitude changes by a factor of 10^{-6} between successive units listed.

4.3

Volume (D)		*V*
Litre (DnC)		l
Cubic metre (DC)		m^3
	m^3	
l	[dm^3]	[L, l][4]), [5])
		[dl]
ml	[cm^3]	[cc, ccm]
μl	mm^3	[λ, ul]
nl		
pl		[μμl, uul]
fl	μm^3	[μ^3, u^3]

Note 1. The litre has been redefined (CGPM, 1964) and is now exactly equal to the cubic decimetre.

Note 2. CQUCC has decided to retain the litre as a permissible unit, especially because of the convenience of its multiples (ml, μl, etc.) and its continued use in chemistry. This decision has become 'respectable' because the International Committee of Weights and Measures (CIPM) has recognized (12) that the litre and a few other units 'play such an important part that they must be retained for general use with the International System of Units'. The preference of litre is supported jointly by a 1972 Recommendation from ICSH, IFCC, and WAPS (6). An eventual choice of 'cubic metre' as preferred unit in practice would not change the significant figures: 1,037 l = 1,037 × 10^{-3} m^3 (= 1,037 dm^3).

Note 3. The 'litre' and the numeral 'one' should be given distinctly different symbols: l and 1 respectively.

Note 4. The order of magnitude changes by a factor of 10^{-9} for each step of the series m^3, mm^3, μm^3.

4.4

Mass (B)	*m*
Kilogram (B)	**kg**
kg	[Kg, K][8])
g	[gr., gm., gms., GR, GRM]
mg	[mgm, mgms]
μg	[γ, ug]
ng	[mμg, mug]
pg	[μμg, uug]

Note 1. The spelling 'kilogram' is recommended (12) in American and English.

Note 2. 'Mass' should not be confused with 'weight' which is a derived kind of quantity. (The dimensions are M and LMT^{-2} respectively.)

4.5

Number (of entities) (D)	*N*
Unity (DC)	1

$\times 10^9, \times 10^6, \times 10^3, \times 1, \times 10^{-3}$

[4]) Symbols in square brackets are not recommended.

[5]) IUPAC and Comité International des Poids et Mesures recently have approved 'L' as an alternative symbol for 'litre'.

Note 1. This kind of quantity has the dimension one (see 3.8 and 3.4, Note 1) and the coherent unit is 'unity' (1).

Note 2. This kind of quantity was called 'number of particles' in QU-R66-4.5.

4.6
Amount of substance (B) *n*
Mole (B) **mol**

mol	[M, *M*, eq, val, g-mol][8])
mmol	[mM, meq, mval]
μmol	[μM, uM, μeq, μval]
nmol	[nM, neq, nval]

Note 1. SCC and IFCC recommend that kinds of quantities of a 'molecular' nature should be used whenever an elementary entity can be stated for a component, be it electrolyte or non-electrolyte, inorganic or organic (see 2.12).

Note 2. 'Amount of substance' supplants the ambiguous concept 'equivalent amount'. The elementary entity (see 2.10) forming the basis of measurement should be indicated whenever doubt may arise. Thus 'Calcium(II) ion(Ca)' would indicate unchelated divalent ionized calcium with oxidation number +II, i.e. $Ca(II)^{2+}$. The component having the entity half as large would be 'Calcium(II) ion($Ca_{0.5}$)' or 'Calcium(II) ion($Ca^{2+}_{0.5}$)' or 'Calcium(II) ion(½Ca)' (see (10), Section 3.6).

Note 3. The unit 'mole' is the amount of substance of a system which contains as many elementary entities as there are atoms in 0,012 kilogram of carbon-12. When the mole is used, the elementary entities must be specified and may be atoms, molecules, ions, electrons, other particles, or specified groups of such particles.

Note 4. The mole is an amount of substance and is not a number of elementary entities. However, the amount of substance constituting one mole contains about $0{,}6022045 \times 10^{24}$ elementary entities.

4.7
Catalytic activity (D) *z*
Katal (special DC) **kat**
(Mole per second (DC)) **(mol/s)**

kkat	(kmol/s)	
kat	(mol/s)	
mkat	(mmol/s)	[kU][8])
μkat	(μmol/s)	[U]
nkat	(nmol/s)	[mU]

Note 1. Catalytic activity of an enzyme is a property measured by the catalysed rate of reaction of a specified chemical reaction, produced in a specified assay system.

Note 2. The catalysed rate of reaction, proposed symbol ξ_{cat}, is defined as the observed rate of reaction, ξ, minus the spontaneous rate of reaction, ξ_0, observed under otherwise similar conditions in the absence of added catalytic material. The rate of reaction (ξ) is defined (10) as the rate of increase of the extent of reaction (ξ)

$$\dot{\xi} = d\xi/dt = \nu_B^{-1} \cdot dn_B/dt$$

where ν_B is the stoichiometric coefficient of component B in the chemical equation for the reaction; ν_B is assigned a negative value if B is a reactant and a positive value if B is a product. The rate of reaction so defined has the same value whichever participant B is selected for measurement (in contrast to the rate of change in the amount of substance of component B itself, dn_B/dt).

Note 3. The catalysed rate of reaction pertains to a reaction in a specified assay system, whereas catalytic activity is considered to be a property of the enzyme, characterizing it even when it is not being tested in the assay system. Catalytic activity is proportional to the (usually unknown) amount of substance of the catalytic component in the material or preparation of interest.

Note 4. Catalytic activity, being defined as equal to (catalysed) rate of reaction, is of the same dimension (NT^{-1}), and is measured in the same SI derived unit 'mole per second'. The special name 'katal' and symbol 'kat' have been proposed by clinical chemists (4,14) and biochemists (2) for this unit in the context of measured catalytic activity. EPQU and Expert Panel on Enzymes have asked IFCC to request CGPM to consider official approval of such use. IUPAC and IUB are expected to support the presentation. CQUCC and EPQU, meanwhile, provisionally recommend that katal and kat be employed in the specific context of catalytic activities measured in medicine.

Note 5. The unit 'katal' is independent of assay conditions, although the measured catalytic activity of a certain system depends on assay conditions. By analogy, the unit metre is constant whereas the length of an iron bar depends on temperature.

Note 6. The term 'catalytic activity' is preferred to such terms in previous use (and somewhat differently defined) as 'catalytic amount' (4) and 'enzymic activity' (2). It should be emphasized that results for catalytic activity imply reference to specified conditions of assay.

Note 7. Values obtained by one method may sometimes be converted to those of another method by means of an *empirical* factor; great caution should be exerted on account of unknown accelerators, inhibitors, and different effects on the individual enzymes in a mixture. In fact, different methods imply different quantities. Conversion is therefore merely an example of the general problem of converting the value of one quantity to that of another by regression analysis.

Note 8. The use of the 'enzyme unit' (symbol U) = 1 μmol/min (9) should be discontinued (2). For any method of assay 1 U = 1 μmol/min = 16,67 nmol/s = 16,67 nkat.

4.8

Mass concentration (D) ρ
Kilogram per litre (DnC) **kg/l**

kg/l	[g/ml][8])
	[%, g%, %(w/v), g/100 ml, g/dl]
g/l	[$^0/_{00}$, g $^0/_{00}$, $^0/_{00}$(w/v)]
	[mg%, mg%(w/v), mg/100 ml, mg/dl]
mg/l	[ppm, p.p.m.(w/v), mg $^0/_{00}$]
	[µg%, µg%(w/v), µg/100 ml, ug/dl]
µg/l	
ng/l	[µµg/ml, uug/ml]

Note 1. This kind of quantity is the mass of the component divided by the volume of the system.

Note 2. The coherent unit is 'kilogram per cubic metre', see 4.3, Note 2. The use of this unit does not change the significant figures or the position of the decimal sign: 0,37 mg/l = 0,37 $\times 10^{-3}$ kg/m^3.

Note 3. Use of different volume bases, e.g. microlitre, millilitre, 100 millilitres, and litre is discouraged. Only the litre should be used.

Note 4. The symbols %, $^0/_{00}$, and p.p.m. are incorrect and ambiguous.

Note 5. Components for which elementary entities can be designated, preferably are measured according to 4.11.

4.9

Mass fraction (D) w
Unity (DC) 1

$\times 1$	[kg/kg]	[g/g][8])
		[%, %(w/w)]
$\times 10^{-3}$	[g/kg]	[$^0/_{00}$, $^0/_{00}$(w/w)]
$\times 10^{-6}$	[mg/kg]	[p.p.m., p.p.m.(w/w), ppm]
$\times 10^{-9}$	[µg/kg]	[p.p.b., p.p.b.(w/w), ppb]
$\times 10^{-12}$	[ng/kg]	

Note 1. This kind of quantity is the mass of the component divided by the mass of the system.

Note 2. This kind of quantity has the dimension one (see 3.8 and 3.4, Note 1) and the coherent unit is 'unity' (1). The unit 'kilogram per kilogram' is a redundant alternative.

Note 3. Owing to the indiscriminate and ambiguous use of per cent (%), per mille ($^0/_{00}$), parts per million (p.p.m.), and parts per billion (p.p.b.) these units should be abandoned.

Note 4. Components for which elementary entities can be designated, preferably are measured according to 4.13.

4.10

Volume fraction (D) φ
Unity (DC) 1

$\times 1$	[l/l]	[ml/ml][8])
		[%, %(v/v), vol%]
$\times 10^{-3}$	[ml/l]	[$^0/_{00}$, $^0/_{00}$(v/v), vol$^0/_{00}$]
$\times 10^{-6}$	[µl/l]	[p.p.m., p.p.m.(v/v), ppm]

Note 1. This kind of quantity is the volume of the isolated component divided by the volume of the system, both volumes at specified conditions.

Note 2. This kind of quantity has the dimension one (see 3.8 and 3.4, Note 1) and the coherent unit is 'unity' (1). The unit 'litre per litre' (or the derived coherent unit 'cubic metre per cubic metre') are redundant alternatives.

Note 3. As 4.9, Note 3.

Note 4. As 4.9, Note 4 and 4.11.

Note 5. Giving dilution 'strength' as 1 : 2, 2 : 3, etc. is ambiguous. The use of 4.10 or a precise description of the procedure is recommended.

4.11

Substance concentration (D) c
Mole per litre (DnC) **mol/l**

mol/l	[M, *M*, eq/l, val/l, N, *N*, n][8])
mmol/l	[mM, m*M*, meq/l, mval/l, etc.]
µmol/l	[µM, uM, µeq/l, etc.]
nmol/l	[nM, neq/l, etc.]

Note 1. This kind of quantity is the amount of substance of the component divided by the volume of the system. The elementary entity of the component should be stated, if doubt may occur.

Note 2. The importance of this kind of quantity should appear from 4.6, Note 1. An alternative is 4.13.

Note 3. Concerning the concept 'equivalent concentration', see the analogous 4.6, Note 2.

Note 4. The derived coherent unit is 'mole per cubic metre', see 4.3, Notes 1–4, and 4.8, Note 2.

Note 5. Concerning the volume base, see 4.8, Note 3.

Note 6. In QU-R66 the name 'molar concentration' was used, but the term 'molar' is now restricted by IUPAC to kinds of quantities having in their definitions 'an extensive kind of quantity of the system divided by amount of substance of the system'. The internationally recommended term for the kind of quantity defined in Note 1 is 'concentration' or 'amount of substance concentration'. After discussions (1971) with IUPAC Commission I.1 it was agreed that CQUCC can recommend the name 'substance concentration' to avoid confusion with the word 'concentration' in the colloquial sense.

4.12

Molality (D) m
Mole per kilogram (DC) **mol/kg**

mol/kg	[m, *m*, mmol/g, µmol/mg][8])
mmol/kg	[mm, m*m*]
µmol/kg	[µm, µ*m*]

Note 1. This kind of quantity is the amount of substance of the component in the solution divided by the *mass* of the *solvent*. The elementary entity of the solute should be stated, if doubt may arise.

Note 2. For another possibility, see 4.13.

Note 3. The kilogram is preferred as denominator in the units.

4.13

Substance fraction (D) or **mole fraction** (D) *x*
Unity (DC) 1

× 1	[mol/mol][8])	
		[%, mol%]
$\times 10^{-3}$	[mmol/mol]	[‰, mol‰]
$\times 10^{-6}$	[μmol/mol]	

Note 1. This kind of quantity is the amount of substance of the component divided by the amount of all components in the system. The elementary entities of the components should be stated, if doubt may arise.

Note 2. This kind of quantity has the dimension one (see 3.8 and 3.4, Note 1) and the derived coherent unit is 'unity' (1). The unit 'mole per mole' is a redundant alternative.

Note 3. As 4.9, Note 3.

Note 4. In QU-R66 the name for this kind of quantity was 'mole fraction', and this, still, is the IUPAC and ISO name; however, the name of a unit does not belong in the kind-of-quantity name and CQUCC and EPQU, therefore, suggest the designation 'substance fraction'. The fully systematic name is 'amount-of-substance fraction'.

4.14

Number concentration (D) *C*
One per litre (DnC) l^{-1} or **1/l**

$\times 10^9$/l	$\times 10^9\ l^{-1}$	
$\times 10^6$/l	$\times 10^6\ l^{-1}$	[× 1/μl, × 1/ul, μl^{-1}][8])
$\times 10^3$/l	$\times 10^3\ l^{-1}$	[× 1/ml, ml^{-1}]
× 1/l	l^{-1}	
$\times 10^{-3}$/l	$\times 10^{-3}\ l^{-}$	

Note 1. This kind of quantity is the number of stated particles or elementary entities of the component divided by the volume of the system.

Note 2. The unit 'one per litre' is preferred to 'one per millilitre' and 'one per microlitre'.

Note 3. The derived coherent unit is 'one per cubic metre', see 4.3, Notes 1–4, and $23 \times 10^3/l = 23 \times 10^6/m^3$.

Note 4. This kind of quantity had the too restrictive name 'particle concentration' in QU-R66-4.14.

4.15

Number fraction (D) δ
Unity (DC) 1

× 1	
	[%][8])
$\times 10^{-3}$	[‰]
$\times 10^{-6}$	

Note 1. This kind of quantity is the number of defined particles or elementary entities constituting the specified component divided by the total number of defined particles in the system.

Note 2. This kind of quantity has the dimension one (see 3.8 and 3.4, Note 1) and the coherent unit is 'unity' (1).

Note 3. As 4.9, Note 3.

Note 4. This kind of quantity had the too restrictive name 'particle fraction' in QU-R66-4.15.

4.16

Catalytic activity concentration (D) *b*
Katal per litre (special DnC) **kat/l**
(Mole per second litre) **(mol · s⁻¹ · l⁻¹)**

kkat/l	$(kmol \cdot s^{-1} \cdot l^{-1})$	[kmol/s/l][8])
kat/l	$(mol \cdot s^{-1} \cdot l^{-1})$	[mol/s/l]
mkat/l	$(mmol \cdot s^{-1} \cdot l^{-1})$	[mmol/s/l]
		[U/ml]
μkat/l	$(\mu mol \cdot s^{-1} \cdot l^{-1})$	
		[U/l]
nkat/l	$(nmol \cdot s^{-1} \cdot l^{-1})$	

Note 1. This kind of quantity is the catalytic activity of the component (enzyme) divided by the volume of the system.

Note 2. The system mentioned in the definition is the original system containing the enzyme, *not* the assay system specified by method of measurement.

Note 3. CNB (2) recommends the name 'concentration of enzyme activity' for this kind of quantity, but has also considered the term 'catalytic activity concentration'. In actual practice, the term 'catalytic activity concentration' may be shortened to 'catalytic concentration'.

Note 4. The coherent SI unit is 'mole per second cubic metre', but CQUCC recommends the use of the special derived non-coherent unit 'katal per litre', in accordance with 4.3, Notes 1–4, 4.7, Note 4, and 4.8, Note 2.
$1\ kat/l = 10^3\ mol \cdot s^{-1} \cdot m^{-3}$.

Note 5. The use of 'enzyme unit per millilitre' (symbol U/ml) should be discontinued, see 4.7, Note 8.
1 U/ml = 16,67 μkat/l.

Note 6. See also 4.7, Note 7.

4.17

Kinds of quantities of a relative nature (D) –
Unity (DC) 1

× 1	
	[%][8])
$\times 10^{-3}$	[‰]
$\times 10^{-6}$	[p.p.m., ppm]
$\times 10^{-9}$	[p.p.b., ppb]

Note 1. Such a kind of quantity is a ratio between a kind of quantity referring to a specified system and an identical kind of quantity referring to another specified 'standard' or 'reference' system.

Note 2. This kind of quantity has the dimension one (see 3.8 and 3.4, Note 1) and the coherent unit is 'unity' (1).

Note 3. As 4.9, Note 3.

Note 4. Examples of this kind of quantity are 'relative length', 'relative substance concentration', and 'relative density' (see 4.25).

4.18
Thermodynamic temperature (B) T
Kelvin (B) **K**
K [°K, °, K°, °C][8])
mK

Note 1. The prior name and symbol (as stated in QU-R66) were 'degree Kelvin' and '°K' respectively. The new forms (kelvin and K) were accepted at CGPM 1967.

Note 2. The unit is defined as the fraction 1/273,16 of the thermodynamic temperature of the triple point of water.

4.19
Celsius temperature (D) θ
Degree Celsius (spec. D) °C
°C, K [C, °, C°, centigrade][8])
m°C, mK

Note 1. The former name 'customary temperature' (as given in QU-R66) was changed to 'Celsius temperature' by CGPM 1967. The symbol, θ, is preferred over t to avoid confusion with the same symbol for 'time'.

Note 2. The unit degree Celsius is identical with kelvin but is used only for Celcius temperatures θ defined by $\theta = T - 273{,}15$ K (exactly), where T is thermodynamic temperature.

Note 3. As 'degree Celsius' is identical to 'kelvin', either may be used to record Celsius temperature, see also 4.20.

4.20
Temperature difference (D) ΔT
Kelvin (DC) **K**
K [°C] [°K, C, °, C°, centigrade][8])
mK [m°C] [mdeg]

Note 1. The unit 'degree', symbolized 'deg' (as given in QU-R66), was abandoned by CGPM 1967.

Note 2. A temperature difference or temperature interval is preferably given in kelvins but can also be given in degrees Celsius.

4.21
Pressure (D) p
Pascal (DC) **Pa**
(Newton per square metre (DC)) (N/m^2)

MPa	(MN/m^2)	
		[atm][8])
		[bar, b]
kPa	(kN/m^2)	
		[mmHg, Torr]
		[mbar, mb]
		[mmH_2O]
Pa	(N/m^2)	[pascal]
		[μbar, μb]

Note 1. This kind of quantity is the force (scalar value) at right angles to a plane divided by the area of that plane. (In physical chemistry, pressure could be defined as the negative value of the partial differential of the internal energy with respect to volume.)

Note 2. IUPAC has decided that the use of the unit 'bar' (1 bar = 10^5 Pa = 10^5 N/m^2) is to be progressively discouraged and CQUCC in 1969 decided that the use of bar (recommended in QU-R66) should no longer be recommended. It should be replaced by the derived SI coherent unit Pa (or N/m^2).

Note 3. Both 'pascal' (Pa) and 'newton per square metre' (N/m^2) are permitted derived coherent units, but Pa is preferred due to simplicity and to avoid duplicate names for the same unit.

Note 4. The unit 'newton' is the force that, when applied to a body having a mass of one kilogram, gives it an acceleration *in vacuo* of one metre per second squared.

Note 5. A measured liquid height requires correction for the temperature of the liquid and the local acceleration of free fall before conversion into any of the units of pressure.

4.22
Partial pressure (D) p
Pascal (DC) **Pa**
(Newton per square metre (DC)) (N/m^2)
See 4.21.

Note 1. This kind of quantity is defined as the product of the substance fraction of the gaseous component and the pressure of the gaseous system. Elementary entities should be given whenever doubt may arise.

Note 2. See 4.21, Notes 2–6.

Note 3. The *tension of a gas in a liquid* is much used by clinical chemists and physiologists and is generally understood as a kind of quantity yielding values equal to the partial pressure of the component in a gas phase in equilibrium with the same component dissolved in the liquid.

4.23
Time (B) t
Second (B) **s**
Minute (DnC) **min**
Hour (DnC) **h**
Day (24 hours) (DnC) **d**
Year (DnC) **a**

	a	[yr.][8])
Ms		
	d	[da.]
	h	[hr.]
ks		
	min	[min., m]
s		[sec., s.]
ms		
μs		[us]

Note 1. The unit 'second' permits a rational use of the decimal system. The unit 'hundredth of a minute' is discouraged; thus 0,15 min should be replaced by 9 s.

Note 2. The designation 'time' should not be used for a specific point in a given calendar; the latter kind of quantity may be called 'calendar time'.

4.24
Mass density (D) ρ
Kilogram per litre (DnC) **kg/l**
kg/l [g/ml][8])
g/l [mg/ml]
mg/l [μg/ml]

Note 1. This kind of quantity is the mass of a system divided by its volume.

Note 2. 'Mass density' should not be confused with 'weight density' ('specific weight') which is the weight of a system divided by its volume.

Note 3. The derived coherent unit is 'kilogram per cubic metre', see 4.8, Note 2.

Note 4. The name 'density' for this kind of quantity is ambiguous because it is also used for 'relative density' (see 4.25) and 'absorbance' (also called 'optical density').

4.25
Relative density (D) *d*
Unity (DC) 1
$\times 1$
$\times 10^{-3}$

Note 1. This kind of quantity is the ratio between the density of the system and the density of a 'reference' or 'standard' system under conditions that should be specified for both systems.

Note 2. This kind of quantity has the dimension one and is an example of the group of kinds of quantities mentioned in 4.17.

Note 3. The term 'specific gravity', often used when water is the reference system, may be confused with 'specific weight'; see 4.24, Note 2.

Note 4. A more rational name for 'relative density' would be 'relative mass density' and this name should be used in case of possible ambiguity.

4.26
Kinds of quantities of an arbitrary nature (non-SI B or D)
International unit (non-SI B or D) **int. unit**
Arbitrary unit (non-SI B or D) **arb. unit**
[iu, u, IU, I.U.][8])

Note 1. A kind of quantity of this nature with corresponding unit may be defined when the result of a measurement is not a part of a recognized kind of quantity having a definable dimension. An indication of method should form an integral part of the name for the kind of quantity. Anyone may define an arbitrary unit; the more authoritative 'international unit' is defined by an international body.

Note 2. Concerning 'titre', see QU-R66, 4.26.3.

Note 3. So-called 'qualitative measurements' and 'semiquantitative measurements' are simply 'quantitative' measurements with restricted modes of expression. The two types may be treated as arbitrary kinds of quantities and the results given as 0 or 1 arbitrary unit and 0, 1, 2, 3, ... arbitrary units respectively. The symbols –, ±, (+), +, ++ etc. or ÷ and ? are not recommended. The possible results may be stated as a specification to the kind of quantity (see Table 6 (7.6), 4.26).

Note 4. Immunoglobulin A is a component for which the arbitrary amount of substance is recorded in international units. The value of a 'Blood–Sedimentation reaction' as arbitrary length (*Westergren* 1926) may be measured in arbitrary units.

5. Printing rules for presentation of data

5.1. **Quantity names.** The *Recommendation 1966* states that a printed quantity name should be unambiguous and contain information on *system* (e.g. Serum), *component* (e.g. Sodium ion), and *kind of quantity* (e.g. substance concentration). The specific form may vary with language and local requirements. Some possibilities will appear from the list of examples in Section 7. For a more thorough discussion consult QU-R66, Parts 6 and 7, and 'List of Quantities. Approved Recommendation 1978'.

5.2. **Printing rules for symbols of units and for numbers.** The internationally recommended rules are given in QU-R66, 3.7.2–5 and 3.8; a few are repeated here.

5.2.1. Symbols for internationally recommended units should be printed in roman (upright) type, remain unaltered in the plural, and should be written without a final full stop. The symbols are in lower case letters, except, when derived from a proper name. *Examples:* The symbol for 'grams' is g, *not* g., gms., or G.

5.2.2. The prefix name or symbol is printed in roman (upright) type as a prefix without a space before the name or symbol of the unmultiplied unit which may not be omitted. Compound prefixes are to be avoided when single prefixes are available; this also applies to the factor in the base unit 'kilogram'. *Examples:* milligram (mg), *not* milli gram (m g); micrometre (μm), *not* micron (μ); picogram (pg), *not* micro-microgram (μμg).

5.2.3. When a unit symbol carries an *exponent*, the factor is raised to the same power as the unmultiplied unit symbol. *Examples:* $1\ \mathrm{mm}^2 = (10^{-3}\ \mathrm{m})^2 = 10^{-6}\ \mathrm{m}^2$, $\mu\mathrm{l}^{-1}$ means $(10^{-6}\ \mathrm{l})^{-1}$, *not* $10^{-6}\ \mathrm{l}^{-1}$.

5.2.4. The prefixes should not be used when the results are given for measurement of quantities having the dimension one, for which the coherent unit is 'unity' (1). *Example:* A mass fraction of 20×10^{-3} or 0.020 should not be written 20 m.

Note. One way of avoiding the bothersome powers of ten would be to create a name for the unit 1.

*5.2.5. **Products*** may be written $ab = a \cdot b = a \times b$, $a^n b^m$; $a^n \times b \times c^p$, etc. *Quotients* should be written

$$\frac{a}{b} = a/b = ab^{-1} = a \times b^{-1};$$

$$ab/c;\ \frac{a/b}{c} = (a/b)/c = a \times b^{-1} \times c^{-1} = \frac{a}{bc}$$

Note. In no case should more than one solidus (/) be employed in a given combination, except when the use of parentheses eliminates ambiguity; e.g. *not* $a/b/c$, but $(a/b)/c = a \times b^{-1} \times c^{-1}$ or $a/(b/c) = a \times b^{-1} \times c$ as the case may be.

5.2.6. The *decimal sign* is a comma on the line (7). In documents in the English language a dot on the line can be used, but the comma is preferred in this Recommendation. If the absolute value of the number is less than one, a zero should precede the decimal sign. Numbers with many digits may be written with small spaces separating groups of three, counting from the decimal sign towards the left and right; a point or comma should not be used for separation. *Example:* 0,602 204 5.

6. Proposed new derived kinds of quantities

6.0. **Development of new kinds of quantities.** In QU-R66, Part 5, some useful kinds of quantities were proposed. Amendments were made by CQUCC in 1966, 1971, 1973, and in 1976, and the results are briefly listed here. The abbreviations for type of kind of quantity and unit are explained in Table 3 (4.0).

6.1

Volume content (D) V_c/m_s
Litre per kilogram (DnC) **l/kg**

Note 1. This kind of quantity is the volume of the (isolated) component (C) divided by the mass of the system (S). *Example:* Patient—Blood, volume content.

Note 2. The term 'content' is under discussion due to ambiguities.

Note 3. 'Volume content' should not be confused with 'specific volume', which is the volume of the system divided by the mass of the system.

6.2

Substance content (D) n_c/m_s
Mole per kilogram (DC) **mol/kg**

Note 1. This kind of quantity is the amount of substance of the component divided by the mass of the system. (This kind of quantity should not be confused with 'molality', see 4.12.) *Example:* Liver—Glucose, substance content.

Note 2. See 6.1, Note 2.

6.2.1

Catalytic activity content (D) z_c/m_s
Katal per kilogram (DC) **kat/kg**

Note 1. This kind of quantity is the catalytic activity of the component divided by the mass of the system.

Note 2. See 6.1, Note 2.

6.3

Mean mass rate (D) $\Delta m/\Delta t$
Kilogram per second (DC) **kg/s**
Kilogram per day (DnC) **kg/d**

Note: This kind of quantity is the mass of the component changed in or moved to or from a system divided by the time during which the component was changed or moved.

6.4

Mean volume rate (D) $\Delta V/\Delta t$
Litre per second (DnC) **l/s**
Litre per day (DnC) **l/d**

Note 1. As 6.3, Note, with substitution of 'volume' for 'mass'.

Note 2. The kind of quantity 'clearance' is usually defined as the product of the substance concentration of the component (or comparable derivative) in the specified output and the volume rate of that output divided by the substance concentration of the component in the specified input. Thus, this methodological definition equals 'mean volume rate' and this name is preferred.

6.5

Mean substance rate (D) $\Delta n/\Delta t$
Mole per second (DC) **mol/s**
Mole per day (DnC) **mol/d**

Note. As 6.3, Note, with substitution of 'amount of substance' for 'mass'. This kind of quantity usually should be preferred for mean mass rate (6.3) or mean volume rate (6.4).

6.6

Mean catalytic activity rate (D) $\Delta z/\Delta t$
Katal per second (DC) **kat/s**

Note. As 6.3, Note, with substitution of 'catalytic activity' for 'mass'. (This kind of quantity should not be confused with 'rate of reaction'.)

6.7

According to the latest views, the definitions given in QU-R66 for 'mass rate', 'volume rate', and 'mole rate' pertain to the kinds of quantities 'mean mass rate', 'mean volume rate', and 'mean substance rate' respectively. In daily practice these last three kinds of quantities are used. The definition of 'mass rate' would be: the rate of change in mass at a specified point in time of the component changed in or moved to or from a system. 'Volume rate' and 'substance rate' are defined analogously.

7. Structure of quantity names

7.1. As mentioned in 5.1, *Recommendation 1966* states that **printed quantity names should contain unambiguous information on system, component, and kind of quantity.** The specific form may vary with language and purpose. The principle is now supported by a joint Recommendation 1972 of ICSH, IFCC, and WAPS (6). For a thorough discussion of structure, see QU-R66, Parts 6 and 7. An abbreviated proposal for English names is given below.

7.2. Lists of names may be given for systems, components, and kinds of quantities respectively. A name for each specific quantity is constructed by selecting an appropriate item from each list. In this document, they are joined together in the following way: The name of the *system*, written with an initial capital letter, is followed by a long hyphen (em dash) –; the name of the *component* also begins with a capital letter, serves for primary alphabetizing, and is followed by a comma; finally comes the name of the *kind of quantity*, written in lower case letters. Names for systems and kinds of quantities may be abbreviated; for kinds of quantities their italicized symbols may also be used.

7.3. **System.** The name of the system, chosen as proper, stands alone. It may be preceded by an explanatory indication of a 'super' system or be followed by a 'sub' system. *Examples:* Blood, (Blood)Leukocytes, Patient(Urine). The full name is unambiguous, but possible codes in English for commonly used systems are given in Table 4 (7.3).

Tab. 4 (7.3). Possible codes for systems (in English).

Code	Meaning	Remark
a	Arterial	(used as a prefix)
B	Blood	
c	Capillary	(used as a prefix)
d	24-hour	(used as a prefix)
Ex	Expectoration	
Erc	Erythrocyte	(or E)
(B)Erc	Erythrocyte (from blood)	(or (B)E)
Ercs	Erythrocytes	(or Es)
fPt	Fasting patient	
F	Faeces	
Lkc	Leukocyte	
Lkcs	Leukocytes	
P	Plasma	
Pt	Patient	
S	Serum	
Sf	Spinal fluid	
U	Urine	
v	Venous	(used as a prefix)

7.4. **Component.** Usually, chemical names should conform with IUPAC nomenclature, whether the names be systematic or trivial. Elementary entities must be appended (in parentheses) when necessary to avoid ambiguity. Enzyme names should conform with IUPAC-IUB Recommendations 1972 (2). Names of methods should not supplant the component name, but, when necessary, should be placed in parentheses as an explanation after the name of the kind of quantity.

Tab. 5 (7.5). Abbreviations for kinds of quantities.

Abbreviation	Meaning
ams.	Amount of substance
arb.	Arbitrary
cata.	Catalytic activity
catc.	Catalytic activity concentration
(cont.	Content)
dens.	Density
diff.	Difference
massc.	Mass concentration
massfr.	Mass fraction
molal.	Molality
substfr.	Substance fraction
numc.	Number concentration
numfr.	Number fraction
rel.	Relative
spec.	Specific
substc.	Substance concentration
temp.	Temperature
vol.	Volume
volfr.	Volume fraction

Abbreviations should be avoided. *Examples:* Nitrogen(N), Calcium(II) ion(Ca), Calcium(II) ion(Ca^{2+})$_{0,5}$ or Calcium(II) ion(positive charge), Fatty acid (carboxyl), Phosphate(P), Chloride, Alanine–meaning the sum of negative, amphoionic, and positive particles–Alanine aminotransferase, Fluid(filtrated), e.g. from the renal glomeruli.

7.5. **Kind of quantity.** The names may be abbreviated as suggested in Table 5 (7.5) or given as their italicized

Tab. 6 (7.6). Examples of systematic quantity names.

Cf. Section	Quantity System–Component, kind of quantity	Reference value num. value unit
4.16	S–Acid phosphatase, catc. (*Jacobsson* 1960)	= 65 nkat/l
4.8	S–Albumin, massc.	= 42 g/l
4.9	(S)Protein–Albumin, massfr. (method)	= 0,69 = 0,69 kg/kg
4.11	S–Albumin(68 000), substc.	= 618 μmol/l
4.11	(fPt)S–Bilirubins, substc.	= 14 μmol/l
4.3	Pt–Blood, vol.	= 4,90 l
4.6	dU–Calcium(II), ams.	= 4,3 mmol
4.11	P–Carbonate + carbon dioxide, substc.	= 28 mmol/l
4.14	Sf–Cells, numc.	= 2 × 10^6/l
4.23	B–Coagulation, time (*Biggs* et al. 1957)	= 1,3 ks
4.17	fPt–Dioxygen(absorb.), rel. subst. rate (rest; Pt/Norm)	= 1,00
4.10	B–Erythrocytes, volfr.	= 0,43
4.26	U–Glucose, arb. substc. (Clinistix®; 0–1)	= 0 arb. unit
4.6	(B)Erc(mean)–Hemoglobin(Fe), ams.*	= 2,0 fmol
4.15	(B)Lkcs–Lymphocytes, numfr.	= 0,3
4.4	Pt–Patient, mass	= 70,8 kg
–	aB–Plasma, pH(37,0 °C)	= 7,39
4.25	B–Serum, rel. dens. (S. 20 °C/H_2O 20 °C)	= 1,026

* Known as 'MCH' or 'mean cell hemoglobin'.

symbols. Necessary specifications are placed in parentheses. *Examples:* relative substance concentration (Pt/Norm), viscosity (4,0 °C), catalytic activity concentration (*Kind & King* 1954).

7.6. **Systematic names.** Examples of quantity names, constructed according to the above rules, are given in Table 6 (7.6). They illustrate the use of different kinds of quantities and are mostly chosen to show a solution to the more difficult problems, rather than the easy. A presumably usual value is given, employing a recommended unit. For more (and often more simple) examples, see QU-R66, 7.3 and List of Quantities. Approved Recommendation 1978.

8. References

(Abbreviations on p. 23)

1. Appendix to this document with list of Correspondence.
2. Commission on Biochemical Nomenclature of the IUPAC and the IUB (1973) Enzyme Nomenclature. Recommendations (1972) on the Nomenclature and Classification of Enzymes together with their Units and the Symbols of Enzyme Kinetics, 443 pp., Elsevier, Amsterdam.
3. Commission on Quantities and Units in Clinical Chemistry of the IUPAC Section of Clinical Chemistry and Expert Panel on Quantities and Units of the IFCC Committee on Standards, Minutes Meeting Copenhagen 1972-06-16 to 17.
4. Dybkær, R. & Jørgensen, K. (1967) Quantities and Units in Clinical Chemistry Including Recommendation 1966 of the Commission on Clinical Chemistry of the IUPAC and of the IFCC, x + 102 pp., Munksgaard, Copenhagen.
5. International Committee for Standardization in Hematology, Implications of IUPAC–IFCC Recommendations for Hematology, 17 pp., unpublished working document.
6. International Committee for Standardization in Hematology, International Federation of Clinical Chemistry and World Association of (Anatomic and Clinical) Pathology Societies: Recommendation for Use of SI in Clinical Laboratory Measurements (1972). (See, e.g., (1973) Z. Klin. Chem. Klin. Biochem. *11*, 93 (only)).
7. International Organization for Standardization (ISO) (1973) General Introduction to ISO 31. General Principles concerning Quantities, Units and Symbols, 14 pp., International Standard ISO 31/0.
8. International Organization for Standardization (ISO) (1973) SI Units and Recommendations for the Use of their Multiples and of certain other Units, 1st edn., iii + 21 pp., ISO 1000.
9. International Union of Biochemistry (1965) Enzyme Nomenclature. Recommendations (1964) on the Nomenclature and Classification of Enzymes, together with their Units and the Symbols of Enzyme Kinetics, 219 pp., Elsevier, Amsterdam.
10. International Union of Pure and Applied Chemistry, Division of Physical Chemistry, Commission on Symbols, Terminology, and Units (1970) Manual of Symbols and Terminology for Physicochemical Quantities and Units, 44 pp., Butterworths, London (also in (1970) Pure Appl. Chem. *21*, 1–44).
11. McGlashan, M. L. (1971) Physicochemical Quantities and Units. The Grammar and Spelling of Physical Chemistry, 2nd ed., x + 117 pp., The Royal Institute of Chemistry, London.
12. National Physical Laboratory (1970) SI. The International System of Units, ix + 45 pp., Her Majesty's Stationery Office, London.
13. IFCC, Committee on Standards, Expert Panel on Enzymes (1975) Provisional Recommendation (1974) on IFCC Methods for the Measurement of Catalytic Concentration of Enzymes. Clin. Chim. Acta *61*, F11–F24, following p. 238 (also in (1975) Z. Klin. Chem. Klin. Biochem. *13*, 471–478; in (1976) Clin. Chem. *22*, 384–391).
14. International Union of Pure and Applied Chemistry, Section on Clinical Chemistry, Commission on Quantities and Units in Clinical Chemistry and International Federation of Clinical Chemistry, Committee on Standards, Expert Panel on Quantities and Units (1974) Quantities and Units in Clinical Chemistry. Recommendation 1973. Pure Appl. Chem. *37*, 517–546.
15. Dybkær, R. (1978) Quantities and Units in Enzymology. Enzyme *22*, 91–123.

Appendix

Bergmeyer, H. U. (Tutzing, Federal Republic of Germany)

British Association of Clinical Biochemists (Leeds, U.K.)

Broughton, P. M. G. (Leeds, U.K.)

Desaty, D. (Rexdale, Ontario, Canada)

Egger, E. (Berlin, German Democratic Republic)

Fischer, W. (Freiburg i. Br., Federal Republic of Germany)

Harnoncourt, K. (Graz, Austria)

Herrmann, R. (Gießen, Federal Republic of Germany)

Holtz, A. H. and *van Assendelft, O. W.* (ICSH Working Party on Quantities and Units) (Bilthoven, The Netherlands)

Jansen, A. P. (Nijmegen, The Netherlands)

Klingmüller, V. (Mannheim, Federal Republic of Germany)

Laue, D. (Cologne, Federal Republic of Germany)

Merten, D. (Jülich, Federal Republic of Germany)

Narayanan, S. (Rutherford, New Jersey, U.S.A.)

New Units in Clinical Chemistry (Dutch Society of Clinical Chemistry) (Bilthoven, The Netherlands)

Rigg, J. C. (Wageningen, The Netherlands)

Riley, W. J. (Perth, Western Australia)

Roth, M. (Geneva, Switzerland)

Schaffer, R. (Washington, D. C., U.S.A.)

Spitzer, R. (Vancouver, BC, Canada)

Tammisto, P. (Majalampi, Finland)

v. Boroviczény, K.-G. (Berlin, Federal Republic of Germany)

Zender, R. (La Chaux-de-Fonds, Switzerland)

List of Quantities in Clinical Chemistry

Prepared for publication by R. Dybkær

Preface

The Commission on Quantities and Units in Clinical Chemistry is a part of the Section on Clinical Chemistry of the International Union of Pure and Applied Chemistry (IUPAC). The Expert Panel on Quantities and Units is a part of the Committee on Standards of the International Federation of Clinical Chemistry (IFCC). These two bodies, the Commission and the Expert Panel, have worked jointly on this document, the former mainly concerned with basic philosophy, the latter with problems of implementation.

The aim of this document is to serve as a guide in supplanting present vernacular names for measurable properties in clinical chemistry. Systematic and more informative names are recommended, based on chemical and biochemical nomenclature. At the same time, 'molecular' kinds of quantities and SI units are preferred.

The tentative version of the present publication appeared as a IUPAC–IFCC (yellow) *Information Bulletin*, Appendices on Tentative Nomenclature, Symbols, Units, and Standards, No. 21, February 1972. As a consequence of comments from knowledgeable colleagues and reconsiderations by the Commission and Expert Panel a Recommendation 1973 contained many, mostly small, modifications from the tentative version. The Recommendation 1973 was approved by IUPAC in 1973 and published in Pure and Applied Chemistry 1974, *37*, 547–572. The IFCC Council in 1975 did not accept the text on account of another view on quantities and units in enzymology presented in the IFCC Expert Panel on Enzymes'

Provisional Recommendation (1974) (6). At a joint meeting in Strasbourg 1976 the Expert Panel on Quantities and Units, the Commission on Quantities and Units in Clinical Chemistry, and the Expert Panel on Enzymes reconciled their concepts, which were further modified according to the advice of the IUPAC Interdivisional Committee on Nomenclature and Symbols. The present document incorporates the resulting changes in names and definitions of kinds of quantities in enzymology.

Where it conflicts, the present document supersedes the larger IUPAC–IFCC Recommendation 1966, its translation into Spanish, and the IUPAC Recommendation 1973.

This Recommendation has been approved by IUPAC, and by the Council of IFCC in February 1978.

Contents

Abbreviations of references

CBN	Commission on Biochemical Nomenclature
CQUCC	Commission on Quantities and Units in Clinical Chemistry
CGPM	Conférence Générale des Poids et Mesures
EPQU	Expert Panel on Quantities and Units
ICSH	International Committee for Standardization in Hematology
IFCC	International Federation of Clinical Chemistry
INN	International Non-Proprietary Name
ISO	Interntional Organization for Standardization
IUB	International Union of Biochemistry
IUPAC	International Union of Pure and Applied Chemistry
IUPAP	International Union of Pure and Applied Physics
NFN	Det nordiske Farmakopénævn
QU-R66	Quantities and Units in Clinical Chemistry. Recommendation 1966 (IUPAC–IFCC) (2)
WAPS	World Association of (Anatomic and Clinical) Pathology Societies

0. Introduction

0.1. In 1967 a monograph (2) (QU-R66) appeared containing a *Recommendation 1966* of the Commission on Clinical Chemistry of the International Union of Pure and Applied Chemistry (IUPAC) and of the International Federation of Clinical Chemistry (IFCC). Since then, the work has been continued by the Commission on Quantities and Units in Clinical Chemistry (CQUCC) of IUPAC and by the Expert Panel on Quantities and Units (EPQU) of IFCC. A condensed and revised version of Qu-R66 was printed as a separate Recommendation (5), now updated (7).

0.2. Concerning names of clinical chemical quantities, the *Recommendation 1966* stated (QU-R66-6.3.1) that *the generic part of the quantity name be unambiguous and contain the following three segments of information:*

system of which a component or function is measured, e.g. Serum

component, e.g. Sodium ion

kind of quantity, e.g. 'molar concentration', now called substance concentration (see 5.2.1).

This statement was supported by a joint Recommendation 1972 from ICSH, IFCC, and WAPS (3).

0.3. On the basis of a thorough discussion (QU-R66-2.3) it was also stated in the *Recommendation 1966* (QU-R66-4.6) that the kind of quantity *amount of substance* is of 'central importance in clinical chemistry'. Consequently, this base kind of quantity (measured in moles) and derived kinds of quantities, e.g. substance concentration and substance fraction (mole fraction), are preferred for mass, mass concentration, mass fraction, respectively. This recommendation was also joint (3).

0.4. These principles, supplemented by some suggested rules, were applied to compile a List of Quantities (QU-R66-7) where the complete systematic names of one hundred quantities were given as examples for discussion.

0.5. The present new list of names is the outcome of further deliberations in CQUCC and EPQU and also incorporates various suggestions from other sources as well as developments in international nomenclature since 1966.

1. Scope of list

The quantities listed are meant to cover many, but not all, of those measured today in clinical chemical laboratories. It is hoped that names for quantities omitted from the list may be created easily by analogy. Not every example is necessarily clinically relevant.

Methods of reporting loading test results, e.g. 'glucose tolerance', are not specifically given. They may contain information on the procedure followed and a table or graph of paired results for time and clinical chemical quantity.

As many clinical chemical laboratories are concerned also with determination of haematological, clinical physiological, clinical microbiological and toxicological quantities, some of these are included.

2. Structure of list

2.1. Each entry is given an alpha-numeric serial reference, e.g. B17, for convenience of identification in this document.

2.2. The general form of an entry is according to *Recommendation 1966* (QU-R66-3.1-1)

quantity = number · unit

2.3. Most of the quantity names listed have the recommended information arranged in the following form

System–Component, kind of quantity

Specifications in parentheses may be necessary to one or more of the three segments. More complex types of names occur, e.g. (A02, A13).

2.4. The quantities are listed alphabetically according to the component that, arbitrarily, has been printed in small capitals.

2.5. The value for each quantity is intended to represent a result from a 'healthy' adult person.

2.6. A set of printing rules is given in *Recommendation 1966* (QU-R66-3.8). The special problems inherent in the use of typewriters with restricted sets of characters have not been treated.

3. Systems

3.1. The system employed is supposed to be a certain 'healthy' person or a part of same (e.g. Blood, Leukocytes, Bilirubins) or some material the person takes up or excretes (e.g. Food, Faeces). Occasionally, a measuring system is mentioned (C13).

3.2. Table 1 (3.2) presents the codes used in this document (for others, see (7)).

Too many abbreviations and codes should be avoided and, in case of doubtful comprehensibility, the full name of the system should be given.

It may be necessary to indicate that the system of the quantity is part of a supersystem: (aB)P means that the plasma of the quantity is derived from arterial blood (D04).

When specifications are necessary, they are placed in parentheses after the system (D03, E04).

Tab. 1 (3.2). Codes for description of system.

Code	Meaning
B	Blood (in vessels)
aB	Arterial blood (in arterial vessels; including 'capillary blood')
vB	Venous blood (in venous vessels)
d	24-hour (change of system during 24 hours)
fPt	Fasting patient (measurement in the morning)
F	Faeces
P	Plasma (blood plasma)
Pt	'Patient' (individual on whom measurement is being made)
S	Serum (blood serum)
Sf	Spinal fluid
U	Urine

3.3. The calendar time of obtaining the specimen should be a part of the identification of the person as well as other relevant information on technique of obtaining the specimen plus the physiological state of the person. That the specimen was obtained in the morning and that the person was fasting is symbolized fPt. Thus, (fPt)S or ((fPt)B)S means that the serum of the quantity is derived from blood of the fasting person in the morning (A17).

The time (interval) during which a registered change occurred in a system is coded in one situation: d for change during 24 hours = 1 d (morning to morning) (A18). Otherwise, the calendar time interval should be specified, e.g. after the kind of quantity.

4. Components

4.1. Names of components should be printed in full.

4.2. The following types of components may be distinguished.

4.2.1. Definable chemical components or *groups of such* are built of characteristic steric structures which may be symbolized by a chemical formula or common elementary entity, e.g. Carbamide, $CO(NH_2)_2$, Amino acid nitrogen, N.

4.2.2. Functional chemical components are chemical compounds or groups of compounds characterized by common properties or effects, e.g. Alkaline phosphatase.

4.2.3. Physical components are composed of microscopic or macroscopic physical bodies, which constitute particles or phases in the system, e.g. Erythrocytes.

4.2.4. Chemical or physiological processes may serve as components, e.g. Coagulation, Capillary bleeding (C34, C09).

4.3.1. Chemical names are in accordance with IUPAC recommendations.

4.3.2. Sodium is the name of the metallic element with atomic number 11 and oxidation number zero. This component is not found in biological systems, whereas *Sodium ion* is present.

4.3.3. If doubt may arise as to oxidation number the *Stock notation* is employed using Roman numerals in parentheses. Thus, Calcium(II) signifies calcium with oxidation number +II irrespective of whether it be free ion Ca^{2+} or different kinds of chelated calcium(II) with ionization charges more difficult to define (C04, C05, C06, C07).

4.3.4. Roman numerals are sometimes used for other kinds of specifications to the component name, e.g. for porphyrins or coagulation factors (U11, C35).

4.3.5. Trivial chemical names have often been chosen for brevity instead of systematic chemical names. Where several trivial names exist for a substance the more 'chemical', i.e. the more informative, is given, e.g. carbamide for urea (C10).

4.3.6. Prefixes defining sterioisomeric structure are omitted if not essential, e.g. Glucose for *D*-Glucose.

4.4. For *drugs*, the names from national or, preferably, international pharmacopoeas should be used (NFN, INN).

4.5. Enzymes are given Recommended Names according to the *Recommendation 1972* from International Union of Biochemistry (1) (A09).

4.6. Blood cells are named according to suggestions by the Working Group on Standardized Documentation of Hematological Findings of the Expert Panel on Hematological Documentation under the International Committee for Standardization in Hematology (ICSH) (B08, E01, P08). This nomenclature is not yet accepted by ICSH, but seems short, logical, and directly understandable.

4.7. Taxonomic names of *genera*, species, and subspecies are printed italic (M16, N01).

4.8. For *acids and bases*, defined according to *Brønsted*, the totality of corresponding acid-base pairs or series are often considered as one group. No accepted rules exist for naming such mixtures. The following rules are recommended:

4.8.1. The designation of the maximally ionized form of those in question is given, omitting the word 'ion'. Thus, Ammonium (A25) comprises ammonium ion and ammonia; Creatininium (C47) comprises creatininium ion and creatinine; Ascorbate (A33) comprises ascorbate ion and ascorbic acid; Carbonate (C18) comprises carbonate ion, hydrogen carbonate ion, and carbonic acid, but *not* carbon dioxide. It should be noted that, in contrast to Ammonia and Ascorbic acid, the designations Ammonium and Ascorbate do not indicate single well-defined, chemical compounds.

4.8.2. Trivial names for organic amphoionic substances having trivial names are used for the totality of the amphoionic, acid, and base forms. E.g. Hydroxyproline (H22) means hydroxyprolinium ion plus hydroxyproline (non-charged) plus hydroxyprolinate ion.

4.8.3. Mixtures of a defined chemical component and its derivatives may be denoted by the plural form of the name of the pure substance (B13, C30).

4.9. The component as specified by a name may have to be further *specified* in parentheses to avoid ambiguity.

The elementary entity (C37, P02) or relative molecular mass (C05, T06) is often necessary information. The latter alternative is especially useful for proteins where the structure is not known in detail, and where the use of mass concepts also needs specification. For mixtures of substances limits may be specified (A17, A18, C35, U11). To indicate the sum of components, specified in individual quantities, the specification 'total' may be applied (C08). Also, specifications relating to a physiological process in which the substance participates may be needed (D05).

4.10. Often several equally correct names are possible for the same component (A21, N08) (C04, C08). Which one is preferred depends on the context.

5. Kinds of quantities[1])

5.1. The kinds of quantities used are those given in *Recommendation 1966* with amendments and additions based on recent IUPAC and IUB recommendations, and approved by CQUCC und EPQU (see (7)).

5.2.1. The English designation 'molar' is now restricted by IUPAC to kinds of quantities with a definition containing the expression 'an extensive kind of quantity of the system divided by amount of substance of the system'. Consequently, 'molar concentration' is no longer acceptable; the systematic name is a cumbersome 'amount of substance concentration' or 'concentration' (4). However, after consultation with IUPAC Commission on Symbols, Terminology, and Units, it was agreed that CQUCC can recommend the name 'substance concentration' to avoid confusion with the word 'concentration' in the colloquial and broader sense.

5.2.2. The IUPAC–IUB Recommendations 1972 (1) contain recent changes in some of the names and definitions for kinds of quantities and units in enzymology. They define the derived kind of quantity 'enzymic activity' as 'the rate of reaction of substrate that may be attributed to catalysis by an enzyme' and have the derived unit 'katal' = 1 mol/s. CQUCC and EPQU in their Recommendation 1973 (5) decided to retain the former suggestion of 'catalytic amount' considered as a base kind of quantity with a base unit 'katal', symbolized kat and defined as the catalytic amount of any catalyst (including any enzyme) that catalyses a reaction rate of one mole per second in an assay system. After consultation with the IUPAC–IUB Joint Commission on Biochemical Nomenclature, the IFCC Expert Panel on Enzymes, and the IUPAC Interdivisional Committee on Nomenclature and Symbols, it was agreed to change the kind of quantity name to 'catalytic activity' and to define this derived kind of quantity as a property of an enzyme 'measured by the catalysed rate of reaction of a specified chemical reaction, produced in a specified assay system' (7). The katal is now proposed by IFCC, IUB, and IUPAC as a special name for the derived unit mole per second, when catalytic activity of an enzyme is involved. The use of the 'enzyme unit' (symbolized U) ≙ 1 μmol/min should be progressively discouraged. For a given method 1 U ≙ 16,67 nkat. The derived kind of quantity 'catalytic activity concentration' (CBN: concentration of enzymic activity) uses the non-coherent unit kat/l; 1 U/ml ≙ 16,67 μkat/l. (The sign ≙ signifies 'corresponding to' and is used instead of an 'equal to' sign when dimensions on the two sides are different.)

5.2.3. CQUCC in Washington, DC, 1971, decided that the kind of quantity 'number of particles' (QU-R66-4.5) be renamed 'number (of entities)'. In accordance, 'particle concentration' (QU-R66-4.14) and the ambiguous 'particle fraction' (QU-R66-4.15) were renamed 'number concentration' and 'number fraction' respectively.

5.3. A few recommended kinds of quantities that have not been systematically described by CQUCC are used in the List with recent modifications.

5.3.1. Volume content (QU-R66-5.1. Specific volume content) is the volume of the isolated component divided by the mass of the system (B16).

5.3.2. Mean substance rate (QU-R66-5.5. Mole rate) is the amount of substance of the component changed in or moved to or from a system divided by the time during which the component was changed or moved (see (D05)).

5.3.3. Mean volume rate (QU-R66-5.4. Volume rate) is defined analogously with 5.3.2, substituting 'volume' for 'amount of substance' (F03).

5.4.1. For so-called *qualitative tests,* a quantitative expression is used, employing a kind of quantity of an arbitrary nature (QU-R66-4.26). The two possible numerical values are given as specification to the kind of quantity: (0–1) (A27). The symbols + and – are not recommended. (See also 5.7.1 and 5.7.3.).

5.4.2. Another, equally acceptable possibility is the use of, e.g., 'substance concentration' without specifications and a result showing whether the quantity is larger than or smaller than the substance concentration indicating the methodological limit of detection, e.g. Urine–Glucose, substance concentration $<$ 12 mmol/l instead of 0.

5.5. So-called *semiquantitative tests* are quantitative and should be treated as such (B14), see also 5.7.3.

5.6.1. In the List, all names of kind of quantity are given *unabbreviated.* If it be necessary to save space, the following, three possibilities may be considered.

5.6.2. The name of the kind of quantity may be *omitted with reference to a local rule, provided that misunderstandings are impossible in practice.* E.g. Plasma–Glucose = 8,2 mmol/l, where 'substance concentration' is omitted after Glucose, would be permissible if the data sheet states that: (i) the kind of quantity employed is given in the local laboratory manual or (ii) all results in 'mmol/l' refer to the kind of quantity 'substance concentration'.

5.6.3. The kind of quantity may be given as its recommended symbol (see QU-R66-4). E.g. Plasma–Glucose,

---

1) ISO, IUPAC, and IUPAP employ the term 'quantity' both for the abstract 'non-addressed' concept of, e.g., 'substance concentration' and for the 'addressed' measurable property of the 'substance concentration of sodium ion in the plasma of a given person at a stated calendar time'.

c = 8,2 mmol/l. All these symbols are in italics (underlined in typewritten text). The meaning of the symbols may be printed on the data sheet or in the laboratory manual.

5.6.4. The name for the kind of quantity is suitably abbreviated, see QU-R66-7.3-2, substituting 'substc.' for 'molc.' as abbreviation for '(amount of) substance concentration'.

5.7.1. Specifications are given in parentheses. The specific form shown is short and should not be taken as a recommendation.

5.7.2. For some kinds of quantities, the method used is a necessary specification: (i) kinds of quantities of an arbitrary nature; (ii) kinds of quantities relating to a component characterized by its function, e.g. an enzyme.

In other quantity names the inclusion or omission of the method is a matter of judgment about the dependence of the results on the current method(s).

5.7.3. If results are restricted to a scale of integers, the closed interval of possible results should be stated, see 5.4.1 and 5.5 (A27, B14).

5.7.4. In the case of quantities involving 'concentration' in liquids, the temperature of the system has been omitted.

5.7.5. In practice, specifications should be adapted to local needs. A number referring to the laboratory manual, which contains all relevant information, is the shortest specification possible.

6. Numbers

6.1. The numbers given in the List (value of quantity divided by unit) serve as illustrations and, presumably, contain only significant figures so that the analytical variation ('error') should influence only the last figure.

6.2. For some 'arbitrary' quantities, the result is always an integer, including 'zero', see 5.4.1 and 5.5 (A27).

6.3. When the magnitude of the result depends on the method used, the number is replaced by a question mark, as for enzymes (A02).

7. Units

7.1. In each case, the size of the unit has been chosen in accordance with the biological level and the estimated analytical variation of the result. Only factors that are 10^{3n}, where n is an integer, should be used.

The units kat, kat/l, and kat/mol are used for enzymes throughout for the reason stated in 6.3, but in most cases sub-units will be required.

7.2. Quantities having the dimension one use the base unit 'unity' (1) and one of the units 10^{3n} is given, e.g. 1 or 10^{-3} (A07, B06).

7.3. As a consequence of recommendations of CGPM (4) the following *deviations from QU-R66* have been made:

7.3.1. 'Thermodynamic temperature' and 'temperature difference' are given in the base unit kelvin (K), not in degree Kelvin (°K) (R01, F08).

7.3.2. 'Pressure' and 'partial pressure' use the unit pascal (Pa), which is identical with 'newton per square metre' (N/m^2) (B17).

8. References

(Abbreviations, see p. 35)

1. Commission on Biochemical Nomenclature of the IUPAC and the IUB (1973) Enzyme Nomenclature, Recommendations (1972) on the Nomenclature and Classification of Enzymes together with their Units and the Symbols of Enzyme Kinetics, 443 pp., Elsevier, Amsterdam.
2. Dybkær, R. & Jørgensen, K. (1967) Quantities and Units in Clinical Chemistry Including Recommendation 1966 of the Commission on Clinical Chemistry of the IUPAC and of the IFCC, x + 102 pp., Munksgaard, Copenhagen.
3. International Committee for Standardization in Hematology, International Federation of Clinical Chemistry and World Association of (Anatomic and Clinical) Pathology Societies: Recommendation for Use of SI in Clinical Laboratory Measurements (1972). (See, e.g., (1973) Z. Klin. Chem. Klin. Biochem. *11*, 93 (only)).
4. International Union of Pure and Applied Chemistry, Division of Physical Chemistry, Commission on Symbols, Terminology, and Units (1970) Manual of Symbols and Terminology for Physicochemical Quantities and Units, 44 pp., Butterworths, London (also in (1970) Pure Appl. Chem. *21*, 1–44).
5. International Union of Pure and Applied Chemistry, Section on Clinical Chemistry, Commission on Quantities and Units and International Federation of Clinical Chemistry, Committee on Standards, Expert Panel on Quantities and Units (1974) Quantities and Units in Clinical Chemistry. Recommendation 1973. Pure Appl. Chem. *37*, 517–546.
6. IFCC, Committee on Standards, Expert Panel on Enzyme (1975) Provisional Recommendation (1974) on IFCC Methods for the Measurement of Catalytic Concentration of Enzymes. Clin. Chim. Acta *61*, F11–F24, following p. 238 (also in (1975) Z. Klin. Chem. Klin. Biochem. *13*, 471–478; in (1976) Clin. Chem. *22*, 384–391).
7. International Union of Pure and Applied Chemistry, Section on Clinical Chemistry, Commission on Quantities and Units in Clinical Chemistry and International Federation of Clinical Chemistry, Committee on Standards, Expert Panel on Quantities and Units (1979) Quantities and Units in Clinical Chemistry. Approved Recommendation (1978). Clin. Chim. Acta *96*, 157F–183F (also in (1979) J. Clin. Chem. Clin. Biochem. *17*, 807–821).

9. Alphabetical list of quantities

Component names in Section 9 are given in small capitals, e.g. ACETO-ACETATE. This does not constitute a recommendation. It is often preferable to use ordinary type or (boldface) heavy type, but with an initial capital, e.g. Acetoacetate or **Acetoacetate**.

The letter l and numeral 1 should have distinct forms in typewritten and printed texts.

Note. The choice between spelling with -æ- or -œ- or -e- is not meant to imply a preference.

Tab. 2 (9). Alphabetical list of quantities (Synonyms are found in Section 10.)

Ser. No.	System–Component(specifications), kind of quantity (specifications) =	Number ·	Unit	Ref.
A01	S–ACETOACETATE, substance concentration	22	μmol/l	
A02	(B)Erythrocytes–ACETYLCHOLINESTERASE, catalytic activity *per* HEMOGLOBIN(Fe), amount of substance	?	kat/mol	
A03	S–ACETYLCHOLINESTERASE, catalytic activity concentration(method)	?	kat/l	
A04	dU–ACID(H), amount of substance (*Jørgensen* 1957)	40	mmol	
A05	S–ACID PHOSPHATASE, catalytic activity concentration(method)	?	kat/l	
A06	S–ACID PHOSPHATASE(TARTRATE SENSITIVE), catalytic activity concentration(method)	?	kat/l	
A07	(dU)Adrenalinium + noradrenalinium–ADRENALINIUM, substance fraction	0,2	1	
A08	dU–ADRENALINIUM + NORADRENALINIUM, amount of substance	0,30	μmol	
A09	S–ALANINE AMINOTRANSFERASE, catalytic activity concentration(method)	?	kat/l	
A10	S–ALBUMIN, mass concentration	42	g/l	A14
A11	Sf–ALBUMIN, mass concentration	0,20	g/l	A15
A12	(S)Protein–ALBUMIN, mass fraction(method)	0,60	1	
A13	S–ALBUMIN, mass *per* GLOBULIN, mass	1,5	1	
A14	S–ALBUMIN(68000), substance concentration	618	μmol/l	A10
A15	Sf–ALBUMIN(68000), substance concentration	2,9	μmol/l	A11
A16	dPt–ALDOSTERONE(PRODUCED), amount of substance(method)	138	nmol	
A17	(fPt)S–ALIPHATIC CARBOXYLATE(C_{10}–C_{26}, NON-ESTERIFIED), substance concentration	0,58	mmol/l	
A18	dF–ALIPHATIC CARBOXYLATE(C_{10}–C_{26}, NON-ESTERIFIED + ESTERS), amount of substance	15	mmol	L11
A19	(fPt)S–ALIPHATIC CARBOXYLATE(C_{10}–C_{26}, NON-ESTERIFIED + ESTERS), substance concentration	11,1	mmol/l	
A20	S–ALKALINE PHOSPHATASE, catalytic activity concentration(method)	?	kat/l	
A21	(fPt)S–AMINO ACID-NITROGEN, substance concentration	3,0	mmol/l	N08
A22	dU–δ-AMINOLAEVULINATE, amount of substance	43	μmol	Note 1
A23	S–δ-AMINOLAEVULINATE, substance concentration	1,2	μmol/l	Note 1
A24	S–AMINOPEPTIDASE(CYTOSOL), catalytic activity concentration(method)	?	kat/l	Note 2
A25	dU–AMMONIUM, amount of substance	51	mmol	
A26	P–AMMONIUM ION, substance concentration	30	μmol/l	
A27	Rectal mucus–AMOEBA, arbitrary number concentration(method; 0–1)	0	arb. unit	
A28	U–α-AMYLASE, arbitrary catalytic activity concentration(method)	?	arb. unit	
A29	S–α-AMYLASE, catalytic activity concentration(method)	?	kat/l	
A30	S–ANION(A^-), substance concentration	155	mmol/l	
A31	S–ANTITHROMBIN III, arbitrary substance concentration(method)	?	arb. unit	
A32	S–α_1-ANTITRYPSIN, arbitrary substance concentration(method)	?	arb. unit	
A33	(fPt)S–ASCORBATE, substance concentration	55	μmol/l	
A34	S–ASPARTATE AMINOTRANSFERASE, catalytic activity concentration(method)	?	kat/l	
B01	U–BACTERIA, arbitrary number concentration(method; 0–1)	0	arb. unit	
B02	U–BACTERIA(LIVING), number concentration(method)	10	10^6/l	
B03	S–BARBITURATES, arbitrary substance concentration(method)	0	arb. unit	Note 3
B04	aB–BASE(H^+-BINDING GROUPS), substance concentration (*Singer & Hastings* 1948)	45	mmol/l	
B05	aB–BASE(H^+-BINDING GROUPS), substance concentration difference(method; Pt–Norm)	0,0	mmol/l	
B06	Erythrocytes–BASOPHIL PUNCTATED ERYTHROCYTES, number fraction	0,20	10^{-3}	
B07	B–BASOPHILOCYTES, number concentration	0,12	10^9/l	
B08	(B)Leukocytes–BASOPHILOCYTES, number fraction	0,01	1	
B09	S–BILIRUBIN, substance concentration	4	μmol/l	
B10	(S)Bilirubins–BILIRUBIN ESTER, substance fraction(method)	0,5	1	
B11	S–BILIRUBIN ESTER, substance concentration	5	μmol/l	
B12	U–BILIRUBINS, arbitrary substance concentration(method; 0–1)	0	arb. unit	
B13	S–BILIRUBINS(NON-ESTERIFIED + ESTERS), substance concentration	9	μmol/l	
B14	F–BLOOD, arbitrary volume fraction(method; 0–4)	0	arb. unit	H06 Note 4

B15	Pt–BLOOD, volume	4,90	l	
B16	Pt–BLOOD, volume content(method)	70	ml/kg	
B17	Pt–BLOOD(ARTERIAL, DIASTOLIC), pressure(method)	10,4	kPa	
B18	Pt–BLOOD(ARTERIAL, SYSTOLIC), pressure(method)	15,6	kPa	
B19	B–BLOOD COAGULUM LYSIS, time(method)	?	s	
B20	S–BROMIDE, substance concentration	0,0	mmol/l	
C01	dF–CALCIUM(II) (Ca), amount of substance	16	mmol	
C02	dU–CALCIUM(II) (Ca), amount of substance	5,2	mmol	
C03	dU,U–CALCIUM(II) (Ca), substance concentration	4,3	mmol/l	
C04	S–CALCIUM(II) (Ca), substance concentration	2,5	mmol/l	C08
C05	S–CALCIUM(II) (Ca, CHELATE, <1000), substance concentration	0,017	mmol/l	
C06	S–CALCIUM(II) ION(Ca, NON-CHELATE), substance concentration	1,3	mmol/l	
C07	S–CALCIUM(II) (Ca, PROTEIN BOUND), substance concentration	1,1	mmol/l	
C08	S–CALCIUM(II) (Ca, TOTAL), substance concentration	2,5	mmol/l	C04
C09	Pt–CAPILLARY BLEEDING, time(method)	0,15	ks	
C10	dU–CARBAMIDE, amount of substance	0,48	mol	
C11	dU–CARBAMIDE, substance concentration	0,40	mol/l	
C12	S–CARBAMIDE, substance concentration	4,1	mmol/l	
C13	Gas(aB equilibrated)–CARBON DIOXIDE, partial pressure(method: 37,0 °C)	5,4	kPa	C14
C14	(aB)P–CARBON DIOXIDE, substance concentration	1,2	mmol/l	C13
C15	Expiratory gas–CARBON DIOXIDE, volume fraction	0,05	1	
C16	(B)Hemoglobin–CARBON MONOXIDE HEMOGLOBIN, substance fraction	0,01	1	
C17	B–CARBON MONOXIDE HEMOGLOBIN(Fe), substance concentration	0,08	mmol/l	
C18	P–CARBONATE + CARBON DIOXIDE, substance concentration	27,5	mmol/l	
C19	U–CASTS, arbitrary number concentration(method; 0–1)	0	arb. unit	
C20	U–CASTS, number (*Addis* 1949; 12 h, night)	5	10^3	
C21	F–CATALASE, arbitrary catalytic activity concentration(method; 0–1)	0	arb. unit	
C22	S–CATION(B^+), substance concentration	155	mmol/l	
C23	Sf–CELLS, number concentration	2	10^6/l	
C24	S–CERULOPLASMIN(150 000), substance concentration	1,6	µmol/l	
C25	dU–CHLORIDE, amount of substance	165	mmol	
C26	dU,U–CHLORIDE, substance concentration	138	mmol/l	
C27	S–CHLORIDE, substance concentration	103	mmol/l	
C28	S–CHOLESTEROL, substance concentration	1,7	mmol/l	
C29	S–CHOLESTEROL(NON-ESTERIFIED), substance concentration	1,7	mmol/l	Note 5
C30	S–CHOLESTEROLS, substance concentration	6,2	mmol/l	
C31	S–CHOLESTERYL ESTER, substance concentration	4,5	mmol/l	
C32	nU–CHORIONGONADOTROPIN, arbitrary substance concentration(method; 0–1)	–	arb. unit	
C33	S–CITRATE, substance concentration	115	µmol/l	
C34	B–COAGULATION, time(method)	?	ks	
C35	P–COAGULATION FACTORS(I + II + VII + X), arbitrary time (*Quick* 1935)	21	arb. unit	
C36	P–COAGULATION FACTORS(II + VII + X), relative arbitrary substance concentration(method; Pt/Norm)	1,00	1	
C37	S–COBALAMIN(Co), substance concentration(method)	350	pmol/l	
C38	S–COPPER(II), substance concentration	18	µmol/l	
C39	dU–COPROPORPHYRINS(I + III), amount of substance	0,23	µmol	

Note 1. The component is also called 4-Oxo-5-aminopentanoate.
Note 2. EC 3.4.11.1; the component was formerly called leucine aminopeptidase.
Note 3. The numerical values equal those of 'S–Barbiturates, substance concentration', but measurement is performed as if the barbiturate(s) present were allylpropymal(NFN), hence 'arbitrary'.
Note 4. The 'concentration' of blood is sought even if its hemoglobin is used as a marker in the method.
Note 5. This type of quantity equals C28, but the former may be used together with C31 for emphasis.

Tab. 2 (Continued).

Ser. No.	System–Component(specifications), kind of quantity (specifications) =	Number ·	Unit	Ref.
C40	P–CORTICOSTERONE, substance concentration	17	nmol/l	
C41	P–CORTISOL, substance concentration	400	nmol/l	
C42	P–CORTISONE, substance concentration	32	nmol/l	
C43	dU–CREATINE, amount of substance	80	μmol	
C44	dU,U–CREATINE, substance concentration	67	μmol/l	
C45	S–CREATINE, substance concentration	40	μmol/l	
C46	S–CREATINE KINASE, catalytic activity concentration(method)	?	kat/l	
C47	dU–CREATININIUM, amount of substance	19,5	mmol	
C48	dU,U–CREATININIUM, substance concentration	16,3	mmol/l	
C49	S–CREATININIUM, substance concentration	0,07	mmol/l	
C50	S–CRYOGLOBULIN, arbitrary substance concentration(method; 0–1)	0	arb. unit	
D01	P–11-DEOXYCORTISOL, substance concentration	8	nmol/l	
D02	dU–2,5-DIHYDROXYPHENYLACETATE, amount of substance	0,0	mmol	
D03	Gas(aB equilibrated)–DIOXYGEN, partial pressure(method; 37,0 °C)	13,0	kPa	D04
D04	(aB)P–DIOXYGEN, substance concentration	0,13	mmol/l	D03
D05	fPt–DIOXYGEN(ABSORBED), relative substance rate(resting; Pt/Norm)	1,00	1	
E01	B–EOSINOPHILOCYTES, number concentration	0,22	10^9/l	
E02	(B)Leukocytes–EOSINOPHILOCYTES, number fraction	0,02	1	
E03	U–EPITHELIAL CELLS, arbitrary number concentration(method; 0–3)	1	arb. unit	
E04	(B)Erythrocyte(mean)–ERYTHROCYTE, diameter	7,5	μm	
E05	(B)Erythrocyte(mean)–ERYTHROCYTE, volume	88	fl	
E06	(B)Erythrocyte(mean)–ERYTHROCYTE LIFE, time	115	d	
E07	U–ERYTHROCYTES, arbitrary number concentration(method; 0–3)	1	arb. unit	
E08	nU–ERYTHROCYTES, number (*Addis* 1949; 12 h, night)	0,7	10^6	
E09	B–ERYTHROCYTES, number concentration	5,00	10^{12}/l	
E10	Sf–ERYTHROCYTES, number concentration	0	10^6/l	
E11	B–ERYTHROCYTES, volume fraction	0,47	1	
E12	dU–ESTRADIOL, amount of substance(female Pt, ovulation peak)	0,035	μmol	
E13	dU–ESTRIOL, amount of substance(female Pt, ovulation peak)	0,082	μmol	
E14	dU–ESTROGEN, amount of substance(postmenopausal Pt)	0,030	μmol	
E15	dU–ESTRONE, amount of substance(male Pt)	0,019	μmol	
E16	B–ETHANOL, substance concentration	0,0	mmol/l	
F01	P–FIBRINOGEN(340000), substance concentration	8,8	μmol/l	
F02	P–FIBRINOLYSIS, time(method)	?	s	
F03	Glomeruli–FLUID(FILTRATED), mean volume rate(creatininium; 1d)	1,6	ml/s	
F04	Tap water–FLUORIDE, substance concentration	40	μmol/l	
F05	(B)Hemoglobin–FOETAL HEMOGLOBIN, substance fraction	0,00	1	
F06	(fPt)S–FOLATES, substance concentration(method)	26	nmol/l	
F07	U–FORMIMINOGLUTAMATE, amount of substance (after *L*-histidine, amount of substance = 97 mmol; 32,4 ks)	100	μmol/l	
F08	S–FREEZING-POINT DEPRESSION, temperature difference	545	mK	
F09	S–FRUCTOSE-BIPHOSPHATE ALDOLASE, catalytic activity concentration(method)	?	kat/l	
G01	(B)Erythrocyte(mean)–GALACTOSE-1-PHOSPHATE URIDYLTRANSFERASE, catalytic activity(method)	?	kat	
G02	S–α_1-GLOBULIN, arbitrary concentration	2,0	arb. unit	
G03	S–α_2-GLOBULIN, arbitrary concentration	5,0	arb. unit	
G04	S–β_1-GLOBULIN, arbitrary concentration	3,0	arb. unit	
G05	S–β_2-GLOBULIN, arbitrary concentration	3,0	arb. unit	
G06	S–γ-GLOBULIN, arbitrary concentration	8,5	arb. unit	
G07	Sf–GLOBULIN, arbitrary substance concentration(method; 0–1)	0	arb. unit	

G08	S–GLOBULIN, mass concentration	22	g/l	
G09	dU–GLUCOSE, amount of substance	2,3	mmol	
G10	U–GLUCOSE, arbitrary substance concentration(Clinistix®; 0–1)	0	arb. unit	
G11	(fPt)P–GLUCOSE, substance concentration	5,6	mmol/l	
G12	Sf–GLUCOSE, substance concentration	2,8	mmol/l	
G13	U–GLUCOSE, substance concentration	1,9	mmol/l	
G14	(B)Erythrocyte(mean)–GLUCOSE-6-PHOSPHATE DEHYDROGENASE, catalytic activity(method)	?	kat	
H01	S–α_2-HAPTOGLOBIN, arbitrary substance concentration(method; 0–3)	2	arb. unit	
H02	S–HAPTOGLOBIN, substance concentration(hemoglobin binding)	7,5	μmol/l	
H03	Pt–height	1,70	m	Note 6
H04	F–HELMINTHES EGGS, arbitrary number concentration(method; 0–3)	0	arb. unit	
H05	(B)Hemoglobin–HEMIGLOBIN, substance fraction	0,01	1	
H06	F–HEMOGLOBIN, arbitrary substance concentration(method; 0–1)	0	arb. unit	B14
H07	U–HEMOGLOBIN, arbitrary substance concentration(method; 0–1)	0	arb. unit	
H08	(B)Erythrocyte(mean)–HEMOGLOBIN(Fe), amount of substance	2,0	fmol	
H09	(B)Erythrocyte(mean)–HEMOGLOBIN(Fe), substance concentration	21	mmol/l	
H10	P–HEMOGLOBIN(Fe), substance concentration	10	μmol/l	
H11	B–HEMOGLOBIN(Fe), substance concentration(ICSH 1966)	9,6	mmol/l	
H12	S–β_2-HEMOPEXIN, arbitrary substance concentration(method; 0–3)	2	arb. unit	
H13	U–HOMOGENTISATE, arbitrary substance concentration(method; 0–1)	0	arb. unit	
H14	P–HYDROGEN CARBONATE ION, substance concentration(blood; $c(O_2)$ = 0,21 mmol/l; $c(CO_2)$ = 1,19 mmol/l; θ = 37 °C)	23,6	mmol/l	
H15	(aB)P–HYDROGEN ION, substance concentration	39,6	nmol/l	P06
H16	P–3-HYDROXYBUTYRATE, substance concentration	10	μmol/l	
H17	S–2-HYDROXYBUTYRATE DEHYDROGENASE, catalytic activity concentration(method)	?	kat/l	
H18	dU–17-HYDROXYCORTICOSTEROID, amount of substance(method)	0,42	mmol	
H19	P–11-HYDROXYCORTICOSTEROID, substance concentration	0,42	μmol/l	
H20	dU–5-HYDROXYINDOLYL ACETATE, amount of substance	30	μmol	
H21	dU–4-HYDROXY-3-METHOXYMANDELATE, amount of substance	25	μmol	
H22	dU–HYDROXYPROLINE, amount of substance	0,16	mmol	
I01	S–IMMUNOGLOBULIN A, arbitrary substance concentration(method)	105	10^3 int. unit/l	
I02	S–IMMUNOGLOBULIN G, arbitrary substance concentration(method)	120	10^3 int. unit/l	G06
I03	S–IMMUNOGLOBULIN M, arbitrary substance concentration(method)	97	10^3 int. unit/l	
I04	S–INSULIN, arbitrary substance concentration(method)	25	10^{-3} int. unit/l	I05
I05	S–INSULIN(5800), substance concentration(method)	183	pmol/l	
I06	S–IODINE(I, PROTEIN BOUND), substance concentration	0,41	μmol/l	
I07	S–ION(A^- + B^+), substance concentration	310	mmol/l	
I08	dU–IRON(II + III), amount of substance	2,3	μmol	
I09	dU,U–IRON(II + III), substance concentration	1,9	μmol/l	
I10	(fPt)S–IRON(II + III)(IN HEMOGLOBIN AND TRANSFERRIN), substance concentration	33	μmol/l	
I11	(fPt)S–IRON(III)(TRANSFERRIN BOUND), substance concentration	23,6	μmol/l	
L01	S–LACTATE, substance concentration	0,50	mmol/l	
L02	S–LACTATE DEHYDROGENASE, catalytic activity concentration(method)	?	kat/l	
L03	dU–LEAD(II), amount of substance	0,07	μmol	
L04	B–LEAD(II), substance concentration	1,2	μmol/l	
L05	dU,U–LEAD(II), substance concentration	0,06	μmol/l	
L06	U–LEUKOCYTES, arbitrary number concentration(method; 0–3)	1	arb. unit	
L07	B–LEUKOCYTES, number concentration	6,5	10^9/l	
L08	Sf–LEUKOCYTES, number concentration	1	10^6/l	
L09	U–LEUKOCYTES + EPITHELIAL CELLS, number(*Addis* 1949; 12h, night)	2,0	10^6	

Note 6. An alternative is 'Pt–PATIENT, height' or 'Pt–BODY, height'.

Tab. 2 (Continued).

Ser. No.	System–Component(specifications), kind of quantity (specifications) =	Number · Unit		Ref.
L10	F–LIPID, arbitrary volume fraction(method; 0–1)	0	arb. unit	
L11	dF–LIPID, mass(method)	4,0	g	A18
L12	(fPt)S–LIPID, mass concentration(method)	6,7	g/l	
L13	S–α_1-LIPOPROTEIN, arbitrary substance concentration(method; 0–3)	2	arb. unit	
L14	S–α_2-LIPOPROTEIN, arbitrary substance concentration(method; 0–3)	2	arb. unit	
L15	S–β-LIPOPROTEIN, arbitrary substance concentration(method; 0–3)	2	arb. unit	
L16	B–LYMPHOCYTES, number concentration	2,0	10^9/l	
L17	(B)Leukocytes–LYMPHOCYTES, number fraction	0,30	1	
M01	S–α_2-MACROGLOBULIN, arbitrary substance concentration(method; 0–3)	2	arb. unit	
M02	dF–MAGNESIUM(II)(Mg), amount of substance	7,2	mmol	
M03	dU–MAGNESIUM(II)(Mg), amount of substance	2,6	mmol	
M04	dU,U–MAGNESIUM(II)(Mg), substance concentration	2,2	mmol/l	
M05	S–MAGNESIUM(II)(Mg), substance concentration	1,0	mmol/l	M08
M06	S–MAGNESIUM(II)(Mg, CHELATE, <1000), substance concentration	0,13	mmol/l	
M07	S–MAGNESIUM(II)(Mg, PROTEIN BOUND), substance concentration	0,30	mmol/l	
M08	S–MAGNESIUM(II)(Mg, TOTAL), substance concentration	1,0	mmol/l	M05
M09	S–MAGNESIUM(II) ION (Mg, NON-CHELATE), substance concentration	0,57	mmol/l	
M10	Pt–mass	70,0	kg	Note 7
M11	U–MELANIN + MELANOGEN, arbitrary substance concentration(method; 0–1)	0	arb. unit	
M12	U–METHYLKETONE, arbitrary substance concentration(Acetest®; 0–1)	0	arb. unit	
M13	B–MONOCYTES, number concentration	0,50	10^9/l	
M14	(B)Leukocytes–MONOCYTES, number fraction	0,05	1	
M15	S–MUCOPROTEIN, mass concentration(method)	0,75	g/l	
M16	Expectorate–*Myobacterium sp.*, arbitrary number concentration(method; 0–1)	0	arb. unit	
N01	Urethral secretion–*Neisseria sp.*, arbitrary number concentration(method; 0–1)	0	arb. unit	
N02	B–NEUTROPHILOCYTES(NON-SEGMENTED), number concentration	0,20	10^9/l	
N03	(B)Leukocytes–NEUTROPHILOCYTES(NON-SEGMENTED), number fraction	0,02	1	
N04	B–NEUTROPHILOCYTES(SEGMENTED), number concentration	6,5	10^9/l	
N05	(B)Leukocytes–NEUTROPHILOCYTES(SEGMENTED), number fraction	0,65	1	
N06	dF–NITROGEN, amount of substance	70	mmol	
N07	dU–NITROGEN, amount of substance	1,10	mol	
N08	(fPt)S–NITROGEN(AMINO ACID), substance concentration	3,0	mmol/l	A21
N09	S–NITROGEN(NON-PROTEIN), substance concentration	19	mmol/l	
O01	S–ORNITHINE CARBAMOYLTRANSFERASE, catalytic activity concentration(method)	?	kat/l	
O02	S–α_1-OROSOMUCOID, arbitrary substance concentration(method; 0–3)	2	arb. unit	
O03	(B)Erythrocytes–OSMOTIC PRESSURE REACTION, arbitrary pressure(method)	140	arb. unit	
O04	dU–OXALATE, amount of substance	0,30	mmol	
O05	dU–17-OXOSTEROID, amount of substance(method)	44	µmol	
O06	(aB)Hemoglobin–OXYHEMOGLOBIN, substance fraction(method)	0,95	1	
P01	U–PHENYLPYRUVATE, arbitrary substance concentration(method; 0–1)	0	arb. unit	
P02	dU–PHOSPHATE(P), amount of substance	35	mmol	
P03	dU,U–PHOSPHATE(P), substance concentration	29	mmol/l	
P04	S–PHOSPHATE(P, NON-ESTERIFIED), substance concentration	1,3	mmol/l	
P05	S–PHOSPHOLIPID(P), substance concentration	2,75	mmol/l	
P06	aB–PLASMA, pH(37,0 °C)	7,41	1	H15
P07	Pt–PLASMA, volume	2,50	1	
P08	B–PLATTULOCYTES, number concentration	225	10^9/l	
P09	U–PORPHOBILINOGEN, arbitrary substance concentration(method; 0–1)	0	arb. unit	

P10	U–PORPHYRINS, arbitrary substance concentration(method)	?	arb. unit	
P11	dU–POTASSIUM ION, amount of substance	69	mmol	
P12	dU,U–POTASSIUM ION, substance concentration	58	mmol/l	
P13	Erythrocytes–POTASSIUM ION, substance concentration	90	mmol/l	
P14	P,S–POTASSIUM ION, substance concentration	4,3	mmol/l	
P15	dU–PREGNANDIOL, amount of substance	10	μmol	
P16	dU–PREGNANTRIOL, amount of substance	12	μmol	
P17	U–PROTEIN, arbitrary substance concentration(Albustix®; 0–1)	0	arb. unit	
P18	U–PROTEIN, arbitrary substance concentration(*Bence Jones* 1948; 0–1)	0	arb. unit	
P19	dU–PROTEIN, mass(method)	60	mg	
P20	S–PROTEIN, mass concentration(method)	71	g/l	
P21	Sf–PROTEIN, mass concentration(method)	0,30	g/l	
P22	U–PROTEIN, mass concentration(method)	0,05	g/l	
P23	Pt–PULSE, frequency(30 s)	1,20	Hz	
R01	Pt–RECTUM, *Celsius* temperature	36,9	°C	
R02	Pt–RECTUM, thermodynamic temperature	310,1	K	
R03	Pt–RESPIRATION, frequency(60 s)	0,20	Hz	
R04	B–RETICULOCYTES, number concentration	30	10^9/l	
R05	(B)Erythrocytes–RETICULOCYTES, number fraction	6	10^{-3}	
S01	U–SACCHARIDE arbitrary substance concentration (*Fehling*; 0–1)	0	arb. unit	
S02	B–SEDIMENTATION REACTION, arbitrary length(method)	5	arb. unit	
S03	dU–SODIUM ION, amount of substance	200	mmol	
S04	dU,U–SODIUM ION, substance concentration	167	mmol/l	
S05	P,S–SODIUM ION, substance concentration	141	mmol/l	
S06	P–SOMATOTROPINE, substance concentration	0,11	nmol/l	
S07	(B)Hemoglobin–SULPHHEMOGLOBIN, substance fraction	0,01	1	
S08	Pt–SULPHOBROMPHTHALEINATE, relative amount of substance(Pt 2,7 ks/Pt 0 ks; method)	0,03	1	
T01	dU–TESTOSTERONE, amount of substance(male Pt)	0,25	μmol	
T02	S–TESTOSTERONE, substance concentration(male Pt)	22	nmol/l	
T03	(fPt)S–THYMOL REACTION, arbitrary substance concentration(*Maclagan* 1944; 0, 1, 2,..)	0	arb. unit	
T04	S–THYROXIN, substance concentration(method)	102	nmol/l	
T05	S–β_1-TRANSFERRIN, arbitrary substance concentration(method; 0–3)	2	arb. unit	
T06	(fPt)S–TRANSFERRIN(74000), substance concentration	30,8	μmol/l	
T07	(fPt)S–TRIACYLGLYCEROL LIPASE(ATOXYL® RESISTANT), catalytic activity concentration(method)	?	kat/l	Note 8
T08	(fPt)S–TRIACYLGLYCEROL LIPASE(QUININIUM RESISTANT), catalytic activity concentration(method)	?	kat/l	Note 8
T09	S–TRIGLYCERIDE, substance concentration(method)	0,75	mmol/l	
T10	(fPt)S–TRIIDOTHYRONINE REACTION, arbitrary substance concentration(method)	?	arb. unit	
T11	F–TRYPSIN, arbitrary catalytic activity concentration(method; 0–1)	0	arb. unit	
U01	dU–URATE, amount of substance	2,90	mmol	
U02	(U)Concrement–URATE, arbitrary substance content(murexide; 0–1)	1	arb. unit	
U03	dU,U–URATE, substance concentration(method)	2,42	mmol/l	
U04	S–URATE, substance concentration(method)	0,28	mmol/l	
U05	Pt–URINE, mass density(20,0 °C)	1,019	kg/l	
U06	Pt–URINE, pH(indicator paper)	6,3	1	
U07	Pt–URINE, relative density(U 20,0 °C/Water 20,0 °C)	1,020	1	
U08	dPt–URINE, volume	1,20	l	
U09	U–UROBILIN, arbitrary substance concentration(method)	10	arb. unit	
U10	(fPt)U–UROBILINOGEN, arbitrary substance concentration(method; 0–1)	0	arb. unit	
U11	dU–UROPORPHYRINS(I + III), amount of substance	10	nmol	

Note 7. An alternative is 'Pt–PATIENT, mass' or 'Pt–BODY, mass'.
Note 8. EC 3.1.1.3; the component (without specification) was formerly called lipase.

10. Index[4])

[4]) Note that abbreviations listed are not recommended (cf. European J. Biochem. *1*, 259–266 (1967)).

II. Expert Panel* on Nomenclature and Principles of Quality Control in Clinical Chemistry

Approved Recommendations on Quality Control in Clinical Chemistry

* J. Büttner, Hannover (DEU), Chairman until 1975–09
R. Borth, Toronto (CAN)
J. H. Boutwell, Atlanta, Ga. (USA)
P. M. G. Broughton, Birmingham (GBR)
R. C. Bowyer, Perth (AUS), Chairman from 1975–09

Part 1. General Principles and Terminology

This paper is based on a Provisional Recommendation (Büttner et al. 1975), which has been revised to take account of the comments received, and has been accepted by the Council of the International Federation of Clinical Chemistry by mail ballot 1978-06-15.

Contents

1. General

1.1

Communities provide funds for clinical chemistry (CC) as a part of health care. The most important purpose of CC is to measure in body fluids and tissues of individual patients, those substances which are relevant for the understanding, prevention, diagnosis, or treatment of diasease.

1.2

This purpose requires many diverse activities. Scope and relevance change as medical knowledge is growing. Measurements, whether of concentration or of any other type of quantity, need a yardstick (standard), a system of units, and a method, as well as personnel and facilities. Interpretation of analytical results must take into account their reliability as well as the variation known to exist in health and disease. Administration, organization, and handling of information are involved at all stages.

1.3

Analytical methods (whether in CC or other fields) are fully described by their performance characteristics: specificity, accuracy, precision, and detectability (jointly referred to as reliability criteria); and speed, cost, technical skill requirements, dependability[1]), and laboratory safety (jointly referred to as practicability criteria). These performance characteristics are both necessary and sufficient: all of them must be considered, and other considerations one might wish to include as relevant to analytical performance can in fact be shown to be covered implicitly by this set of concepts. Some of the performance characteristics are closely related as to be synonymous or entirely determined by each other. For example, while poor specificity is likely to produce inaccuracy, it may happen that loss or interference causes considerable inaccuracy in a very specific method; or, while a low detection limit will often go together with low imprecision, a very precise method may have a high detection limit due to high blank values. If the degree of reliability achievable with a method is less than required for a stated purpose, the method should be declared unsuitable.

---

1) This concept refers to the risk or frequency of breakdowns.

1.4
The definition of the term Quality Control (QC) in CC deserves careful consideration because of the special technical meaning involved.

1.4.1. In its widest sense, QC would have to be concerned with the reliability of laboratory information about patients. This information would include not only laboratory results, but also the background data which are needed to interpret results correctly, such as circadian and other intra-human variation, the effect of drugs, and knowledge of the appropriate normal range or reference values. However, most of these factors are outside the control of the laboratory.

1.4.2. Therefore, QC in CC is defined (see Section 5 for definitions of technical terms) as the study of those sources of variation which are the responsibility of the laboratory, and of the procedures used to recognize and minimize them, including all sources (such as random variation and bias) which arise within the laboratory between the receipt of the specimen and the despatch of the report. The responsibility of the laboratory may extend to the collection of the specimen from the patient, and the provision of a suitable container. Crude errors or mistakes (e.g. wrong sample identification, incorrect transcription), although they cannot be controlled by the usual statistical QC methods, are nevertheless the concern of the laboratory.

1.4.3. In a narrower sense, the term QC in CC refers mainly to the monitoring of precision and accuracy of the performance of analytical methods.

1.4.4. It must be made clear to laboratory personnel that QC is an obligation to the patient, that it is designed to give analysts confidence in the methods used, and that its purpose is not to find scapegoats or to punish mistakes. There should be clear instructions and assignments of responsibility for remedial action when the performance of an analytical method deteriorates.

1.5
Different authors have adopted different basic philosophical positions (different models of reality) in discussing the "true meaning" of the terms and concepts used in QC. It can be argued, for example, that there is no single true value for a quantity, but rather a distribution of true values. Others consider that the true value is unknowable. The choice of a particular position, although endlessly debatable, is essentially arbitrary. Certain basic assumptions are implicit in any treatment of the subject, and the following premises are adopted in this document:

(1) Concentration and other types of quantity in CC are continuous variates in the statistical sense.
(2) Specimen quantities have true values.
(3) The purpose of measurements is to obtain results which are useful estimates of true values.
(4) Precision and accuracy refer to a set of results obtained under stated conditions, not to single results.
(5) Probability is a relative frequency of change occurrence derived from stated assumptions.

1.6
For important practical reasons, such as problems of operation and experimental design, it is convenient to make a distinction between internal QC (within a laboratory), and external QC (involving at least two laboratories). The theoretical basis is the same.

1.7
Essentially, both precision and accuracy are assessed by doing replicate assays on samples of the same specimen which are assumed to be stable and identical regarding the quantity investigated. In statistical terminology, the replicate results used are considered a random sample from a hypothetical population of replicates; the mean (as well as other measures of location or central tendency) of this sample reflects the performance of the method with respect to accuracy, and the standard deviation (S.D.) (as well as other measures of dispersion) with respect to precision.

1.8
Both precision and accuracy are conventionally measured in terms of quantities which decrease in value as the reliability is increased or improved: more precise methods have smaller standard deviations between replicate results; more accuracy is reflected in a smaller deviation of results from the true value. There is some linguistic awkwardness in this inverse relationship. In the past, this has been dealt with in three ways:

(1) Most authors have ignored it.
(2) Others have defined new parameters which are reciprocals of the conventional ones; e.g. the inverse value of the S.D. has been called "precision".
(3) Still others have used the dictionary words imprecision and inaccuracy in their respective ordinary meanings to rename the conventional parameters, that is, the S.D. has been called *imprecision* and the deviation from the true value *inaccuracy*.

In this document, the third approach has been selected as the preferred compromise between linguistic consistency and current usage. The exact definitions are given in Section 5. It is realized that the compromise may satisfy neither the advocates of the two other possibilities nor those who would prefer other terms for the same concepts.

1.9
This part of the recommendations is an attempt to produce consistency and unification of concepts and terms through selection from the literature combined with some innovation. To achieve wide acceptability and rapid general adoption in practice, it was thought mandatory that innovations should be limited in scope, easily comprehensible, and recognizably derived from tradition and current practice.

2. Precision

2.1 According to the definition adopted (Section 5), precision should be understood to have no numerical value, and the measured parameter corresponding to his performance characteristic should be referred to as imprecision; roughly equivalent terms sometimes used in the past include "scatter" and "analytical variability". Imprecision is determined by making replicate assays of the same specimen, or of each of a set of specimens, according to a stated replication design (e.g. between-days, within-series, etc.), and then calculating the S.D. of replicate results per specimen. The specimens used may be patients' specimens received for analysis, or they may be control specimens. The only assumption, whether tested or tacit, is that the replicate samples of a specimen have the same true value.

2.2

The S.D. obtained in this way may reflect only unattributed random variation, or it may include identifiable sources of variation arising from the use of different specimens, concentrations, series, reagents, instruments, analysts, days (or other periods of time), methods, and laboratories. Each of these may be of methodological interest, and each requires an adequate experimental design and statistical analysis for its determination. These factors can contribute to imprecision and, very important, can also produce inaccuracy. In reporting estimates of imprecision, it should always be clearly stated which sources of variation have been included in the experimental design used.

2.2.1. In QC in industry and in QC of the performance of measuring instruments, it is apparently useful to define two different terms for the precision determined under two selected sets of extreme conditions, "Repeatability" is used to refer to tests performed at short intervals in one laboratory, by one operator using the same equipment, while the term "reproducibility" refers to tests performed in different laboratories. These terms are not likely to be of much use in CC, where it is often necessary to assess imprecision for several different sets of the many contributing factors listed in 2.2, while it is rarely of interest to single out conditions which could correspond to those two definitions.

2.3 In the absence of recognized criteria for cost-benefit analysis, recommendations regarding the frequency and design of replicate assays for QC purposes have to be based on the intuitive assessment, by experienced workers, of minimum requirements and maximum feasibility. Since assay results within a laboratory will usually be subject to most of the sources of variation listed in 2.2, it seems reasonable to include them all in the QC estimation of precision.

2.4 Precision is said to be under control as long as the replicate variation (imprecision) remains below a limit predicted from probability considerations and previous performance. The probability level chosen is the permissible frequency of values exceeding this limit.

2.5

The precision of results obtained by an analytical method may be under control as defined in 2.4 without being satisfactory for certain purposes. No simple rigorous criterion for acceptability of a method is likely to cover even the most obvious purposes in CC, such as the detection of single abnormal values, of a treatment effect, or of a time trend. The usefulness of an analytical method in CC is established through a slow process of trial and error, and the relation between achievable precision, biological variation, and the magnitude of changes to be detected is part of this empirical assessment. Consequently, there are at present three schools of thought. One school regards imprecision as an descriptive parameter reflecting the state of the art, which is subject to almost limitless improvement. Another school recommends that imprecision should in some way be related to biological variation such as that between and within healthy individuals. The third school stipulates pragmatically that the precision of an analytical method should be assessed in terms of its clinical usefulness.

3. Accuracy

3.1

The accuracy of analytical results will be affected by the same identifiable and non-identifiable sources of variation which contribute to imprecision (Section 2.2). Each may be of methodological interest, and each can be investigated using an adequate experimental design and statistical analysis. In addition, the calibration process and calibration standards can be important sources of inaccuracy. These problems require careful consideration before methods of monitoring inaccuracy can be discussed usefully.

3.2 Calibration standards in clinical chemistry

3.2.1. In some contexts, the use of the word standard, without additional explanation, may be ambiguous, e.g. standards for instruments (which could refer to calibration standards or to performance requirements), standardization of methods (which could refer to their calibration or their uniformity). Whenever necessary to avoid such ambiguity, the specific term *calibration standard* should be used for the standards discussed in this section.

3.2.2. It is customary to set up one or more calibration standards with each run or series of samples, and to calculate the assay results from the calibration readings. This approach has several important limitations.

3.2.2.1. The method may not be specific: the sample then generates an unwanted signal which does not arise with a pure substance. In an attempt to overcome this, the substance may be incorporated into a biological matrix so that the calibration standard resembles the sample. However, the non-analyte constituents may be different in standards and samples and may vary between samples. Consequently, non-specific methods will introduce varying inaccuracies with different samples which cannot be corrected by changing the calibration material.

3.2.2.2. With some analyses, it may be impossible to put a calibration standard through all stages in the analytical procedure.

(i) Sometimes aqueous standards cannot be treated in exactly the same manner as specimens (e.g. methods involving protein precipitation or enzymic degradation of conjugates).

(ii) The calibration standard may be unstable, expensive, or inadequate for other reasons, and therefore not easily used in daily routine analysis. In these cases, the specimen concentration may be calculated by using some characteristic of the standard (e.g. molar absorption coefficient) or by comparison with some other material which has a known relationship with the standard.

(iii) Calibration standards for the analyte may not exist (e.g. enzymes). In such cases, the result is calculated partly or solely from instrument readings of time, temperature, mass, volume, pressure, etc.

3.2.3. In contrast to the historical definition of five grades of standards by IUPAC (26), standards in CC are most conveniently classified into two easily recognizable groups. Solutions only are considered here, but the approach is valid for gases and solids as well.

(i) A primary standard solution is one in which the concentration is determined solely by dissolving a weighed amount of standard material in an appropriate solvent and making to a stated volume or mass. The accuracy of a primary standard solution depends solely on the purity of the standard material and solvent, and the accuracy with which the solution is prepared.

(ii) A secondary standard solution is one in which the concentration or other quantity is determined by an analytical method of stated reliability. The accuracy of a secondary standard therefore depends on the accuracy of the analysis, which in turn will involve a standard: ultimately this must be a primary standard. With this classification, both potassium permanganate solution and serum can function as secondary standards. In this document, no attempt is made to distinguish between them by use of terms such as calibration material, reference material, or standard specimen.

3.2.4. The concentration or other type of quantity of the standard must be stated in approved units. This value may be a stated value, reference method value, or certified value (see Section 5).

3.2.5. In CC a wide variety of calibration standards may be used, and these may be classified by using the following additional descriptive terms as appropriate:

(i) Composition, e.g. nature and purity of constituents and of any solvent or matrix.

(ii) Physical state, e.g. solid, solid after lyophilization, liquid, solution, suspension, gas.

(iii) Origin, e.g. name of manufacturer, batch number, species, tissue.

(iv) Quantification, e.g. determined by weighing (primary standard) or analysis (secondary standard), together with a statement of confidence limits. For secondary standards, the method by which the value is obtained must be stated, together with any assumptions made.

(v) Stability, e.g. shelf life under stated storage conditions, nature and amount of any preservatives.

(vi) Application, e.g. the type of analysis for which the standard is intended.

(vii) Additional terms: in some instances it may be useful to add additional terms, such as international, arbitrary, internal, etc. These must always be defined.

3.3

Various methods have been proposed to monitor accuracy: the use of control specimens, recovery experiments, calculation of the daily mean of patients results, and the assay of specimens obtained from healthy subjects. The methods differ considerably in their theoretical basis and applicability.

3.3.1. Inaccuracy can be monitored by assaying control samples assumed to have the same true value, namely, that of the control specimen from which they are taken. If the true value of the control material is not known, the procedure will measure only changes in inaccuracy. If the true value of the control material is not known, the procedure will measure only changes in inaccuracy. If the true value is known, the procedure will, in addition, measure the inaccuracy, i.e. the mean deviation from the true value in a set of replicate results. Calibration and blank readings should be monitored along with results of control assays, as accuracy changes detected by the control system can often be traced to changes in the calibration process.

3.3.2. In recovery experiments, assays are made on a pair of samples of a specimen, a known amount of substance having been added to one of the two samples. The difference between results within the pair is the amount recovered, and the deviation of this from the amount added is a measure of the inaccuracy in the range covered and for the sample used.

3.3.3. According to the daily mean method, the mean is calculated from results obtained during a day, or other suitable period of time. Within such a period, either all results are used, or only those which fall within a suitable defined interval; the definition of the latter may be quite arbitrary. Changes in this mean are taken to reflect changes in accuracy. This claim is based on the premise that day-to-day changes in the set of specimens received for analysis are negligible in their effect on the mean.

Whether this is true or not will depend on local conditions and their change with time. It is not surprising, therefore, that confirmatory as well as refuting evidence on the usefulness of this method has been published. It should be noted that the alleged stability of the hospital population mean can be properly investigated only by an independent method of detecting inaccuracy, and that apparent stability of this mean may possibly mask true changes in accuracy. On the other hand, it is noteworthy that the daily mean method is the only QC method which takes into account sources of variation which might affect specimens before they arrive in the laboratory.

3.3.4. When introducing a new method, the laboratory will usually determine a series of reference values in apparently healthy subjects ("normal range"). Any subsequent alteration in the accuracy of the method will produce a change in reference values. If the reference interval in healthy subjects is small, one or more reference specimens (e.g. from laboratory staff) can be used as control specimens to monitor changes in accuracy. This method of QC is particularly useful for components which are absent or unstable in other types of control specimens.

3.4

There are no agreed criteria for cost-benefit analysis, and the frequency and design of control assays will be decided on the basis of an intuitive assessment of requirements and feasibility (as in Section 2.3). If these assays are done in replicate, they will provide data on precision as well as accuracy.

3.5

Inaccuracy should be zero within the error of the estimate. If it is not, remedial steps should be taken to make it so. Circumstances, including the state of the art, may make inevitable the temporary use of an inaccurate method. Continued toleration of inaccuracy is definitely not acceptable in CC.

3.6

Assay results can be numerically corrected for known inaccuracy. However, it should be emphasized that this procedure is valid only under the assumptions that the inaccuracy (i) has been reliably determined, and (ii) is known to vary no more than negligibly between specimens. Unless these assumptions are verified, routine numerical correction may produce a deceptive impression of good accuracy not warranted by fact.

4. Statistical methods

4.1

Statistical methods are an essential part of QC. Specimens of control material are included in a run of patients' specimens, and the results found for the control specimens are regarded as representative of all of the patients' specimens in that run. Consequently, there is always the implicit assumption and intent that the control specimens behave and are treated in the same way as patients' specimens. The results of the control specimens undergo statistical tests which are intended to show whether the analytical method is "under control". Before using a statistical test, the applicability of the mathematical model from which it is derived should be examined; for example, the distribution of the data should be approximately Gaussian if the test requires this for its validity. From past experience it seems likely that means and standard deviations will be adequately applicable to most QC problems in CC, and that recourse to other parameters (e.g. median, mode, mean deviation) or to non-parametric statistical methods will seldom be of significant advantage.

4.2

The main statistical tool employed in QC is the control chart. This is a graph in which the quantity or statistic to be controlled is plotted against time of observation. The variation of the quantity or statistic with time is estimated, preferably with at least 20–30 degrees of freedom, during an initial period in which the performance of the analytical method was regarded as satisfactory, either intuitively, or by comparison with objective criteria when these exist. From the observed distribution, control limits are calculated within which most (customarily, 95 or 99 per cent) of the values are likely to fall, and these limits are also plotted on the graph. The method is said to be under control as long as values outside the control limits occur less frequently than once in 20 or 100 times, depending upon the chosen probability of chance occurrence. The calculation of control limits is repeated at suitable intervals, based on the performance data which accumulate over months and years of satisfactory operation. Control charts should be kept up to date and displayed.

4.3

Occasionally, an observed change may be just large enough to be significant in terms of the statistical test used, while being at the same time so small that it can be disregarded for all practical purposes. Statistical tests should not be applied mechanically: while a statistically significant change is, in terms of the statistical model used, unlikely to have occurred by chance, it remains nevertheless the analyst's (not the statistician's) responsibility to decide what is relevant in the practical context. This general rule applies to QC as it does to all other use of statistical tests.

4.4

Conversely, from a mechanical use of the control chart, the risk for committing an "error of the second kind", i.e. of failing to detect a real change in performance, may seem to be of little importance in QC, since it is the logical basis of the control chart method that fluctuations, whether real or random, are negligible as long as they remain within the control limits. In practice, however, a consistent trend or permanent change should, of course, not be ignored by the analyst merely because control limits fail to be crossed.

4.5

If markedly discrepant results, so-called outliers, are discarded and replicates averaged, variation is, of course, reduced, and the control chart will look better (despite the fact that calculated limits will also be narrower). Some authors seem to imply that this is an end in itself, and they advocate these procedures as if they were means to improve performance. Needless to say this is not so, and the frequency of outliers and the magnitude of run-to-run variation must not be ignored in assessing performance.

4.6

The quantity plotted may be a single result on a control sample or a mean (inaccuracy), a standard deviation or range (imprecision). The underlying distribution may or may not be approximately Gaussian, and the statistical methods required to calculate the percentiles used as control limits must be chosen accordingly.

4.7

A useful refinement in plotting accuracy data is the so-called cumulative sum ("cusum") chart, which enhances the visibility of changes and trends. The statistic plotted is the accumulating algebraic sum of signed, i.e. positive or negative deviations of the results from their expected value. Accuracy is said to be under control as long as the plotted points scatter randomly about a line of constant slope. This is assessed by testing randomness or slope variation. It should be emphasized that the slope of the plotted line cannot be zero unless the expected value used is exactly equal to the mean value found.

4.8

Appropriate designs and analysis of variance techniques will provide information on the relative importance of sources of variation. The calculation of variance components may be of interest, but a single component should, of course, not be used to predict variation which in reality is bound to include several components.

5. Terminology

Terms are listed in alphabetical order. Cross references to terms which are defined elsewhere in the list are indicated by *italics.* Undefined terms are used in their oridinary dictionary meanings which are assumed to be obvious from the context. The list includes some terms which will be used in subsequent parts of the Recommendations.

Accuracy. Agreement between the mean estimate of a quantity and its *true value.* See *inaccuracy.*

Aliquot. Measured portion of a whole having the same composition. General term referring to any solution, sample, mixture, etc.

Analyte. The component to be measured.

Analytical method. See *Method, analytical.*

Analytical range. See *range, analytical.*

Assigned value. See *value, assigned.*

Bias. Same as *inaccuracy.*

Calibration. Procedure of relating the *reading* to the quantity required to be measured.

Carryover. The influence of a sample upon a following one.

Certified value. See *value, certified.*

Control material. Material used for quality control purposes.

Control specimen, control solution. Specimen or solution which is analyzed solely for quality control purposes, not for calibration.

Definitive method. See *method, definitive.*

Definitive value. See *value, definitive.*

Detectability (often referred to as *sensitivity,* which is used in a different meaning in other fields). The ability of an analytical method to detect small quantities of the component. It has no numerical value. See *detection limit.*

Detection limit. The smallest single result which, with a stated probability, can be distinguished from a suitable blank. The limit may be a concentration or an amount and defines the point at which the analysis becomes just feasible. See: Definition given by IUPAC, Analytical Chemistry Division (20).

Error. Difference between a single estimate of a quantity and its *true value.* This difference or deviation (positive or negative) may be expressed either in the units in which the quantity is measured, or as a percentage of the true value. If a good estimate of the true value is not available, the difference may have to be expressed as the deviation from an assigned value. Note that errors as defined cannot be classified as "random" or "systematic" without considering sets of measurements (see *imprecision, inaccuracy*).

External quality control. Procedure of utilizing, for *quality control* purposes, the results of several laboratories which analyze the same specimen(s).

Imprecision. Standard deviation or coefficient of variation of the results in a set of replicate measurements. The mean value and number of replicates must be stated, and the design used must be described in such a way that other workers can repeat it. This is particularly important whenever a specific term is used to denote a particular type of imprecision, such as between-laboratory, within-day, or between-day.

Inaccuracy. Numerical difference between the mean of a set of replicate measurements and the *true value.* This difference (positive or negative) may be expressed in the units in which the quantity is measured, or as a percentage of the true value.

Interference. The effect of a component, which does not by itself produce a reading, on the accuracy of measurement of another component.

Internal quality control. Procedure of utilizing the results of only one laboratory for *quality control* purposes.

Method, analytical. Set of written instructions which describe the procedure, materials, and equipment, which are necessary for the analyst to obtain a result.

Method, definitive. A method which after exhaustive investigation is found to have no known source of inaccuracy or ambiguity.

Method, reference. A method which after exhaustive investigation has been shown to have negligible inaccuracy in comparison with its imprecision.

Method of known bias. An analytical method of unknown accuracy.

Precision. Agreement between replicate measurements. It has no numerical value. See *imprecision.*

Quality control in clinical chemistry. The study of those errors which are the responsibility of the laboratory, and the procedures used to recognize and minimize them. This study includes all errors arising within the laboratory between the receipt of the specimen and the despatch of the report. On some occassions, the responsibility of the laboratory may extend to the collection of the specimen from the patient, and the provision of a suitable container. See *external* and *internal quality control.*

Random. Due to unattributed causes; not affected by any kind of systematic selection or arrangement.

Range, analytical. The range of concentration or other quantity in the specimen over which the method is applicable without modification.

Reading. The value indicated on the scale of an instrument or analytical device.

Reference method. See *method, reference.*

Reference method value. See *value, reference method.*

Reference value. Quantity with which an observed value is compared for interpretative purposes (this definition has been proposed by the IFCC Expert Panel on Theory of Reference Values).

Result. Final value obtained for a measured quantity after performing a measuring procedure including all sub-procedures and laboratory evaluations.

Run. This usually refers to a set of consecutive assays performed without interruption. The results are usually calculated from the same set of calibration standard readings. However, this definition may not be universally applicable, and in these cases the word *series* should be used after defining it.

Sample. The appropriately representative part of a specimen which is used in the analysis. This sample should be called a test sample when it is necessary to avoid confusion with the statistical term "random sample from a population".

Sensitivity. See *detectability.*

Specificity. The ability of an analytical method to determine solely the component(s) it purports to measure. It has no numerical value. It is assessed on the evidence available on the components which contribute to the result, and on the extent to which they do.

Specimen. Material available for analysis.

Standard. Material or solution with which the sample is compared in order to determine the concentration or other quantity. The compound term *calibration standard* (sometimes referred to as "calibrator") should be used whenever needed to avoid confusion with other technical or colloquial meanings of the word standard.

Standard, arbitrary. Calibration standard containing an unknown quantity of specified substance. The content is assigned by convention for the material concerned. Examples: international biological standards, immunoglobulin standards.

Standard, internal. A substance, not normally present in the specimen and clearly distinguishable from the analyte, which is added in known amount to the sample, or to both the standard and the sample, for the purpose of correcting results for inaccuracy.

Standard material, primary. Substance of known chemical composition and sufficient purity to be used in preparing a *primary standard solution.*

Standard solution, primary. Solution used as calibration standard in which the concentration is determined solely by dissolving a weighed amount of primary standard material in an appropriate solvent, and making to a stated volume or weight.

Standard solution, secondary. Solution used as a calibration standard in which the concentration or other quantity has been determined by an analytical method of stated reliability.

Stated value. See *value, stated.*

Values. (of concentration or other types of quantity in calibration standards, control specimens, etc., as used in the following compound terms; the method by which the value is obtained must always be stated).

Value, assigned. Value assigned either arbitrarily (e.g. by convention), or from preliminary evidence (e.g. in the absence of a recognized *reference method*).

Value, certified. Value certified by some official body subject to conditions established by that body.

Value, certified. Value certified by some official body subject to conditions established by that body.

Value, definitive. A value given by a definitive method. Best available estimate of the *true value.*

Value, definitive. A value given by a definitive method. Best available estimate of the *true value.*

Value, reference method. Most probable value derived from a set of results obtained by the most reliable *reference methods* available.

Value, stated. Value without official certification.

Value, true. Term considered to have a self-evident meaning requiring no definition. In practice the true value is approximated by the *definitive value*, and somewhat less closely by the *reference method value.*

Comment: The definitions given for the various *values* correspond to the definitions of methods as follows:

Method	Value
–	True value
Definitive method	Definitive value
Reference method	Reference method value (stated or certified)*
Method with known bias Method with unknown bias	Assigned value (stated or certified)*

* These four definitions reflect two independent classifications and are, therefore, not exclusive: reference method values as well as assigned values may at the same time be either certified or stated values.

6. References

1. Anido, G., van Kampen, E. J. & Rosalki, S. B. (eds.) (1973) Progress in Quality Control in Clinical Chemistry. Transactions of the 5 th International Symposium, Geneva, April 10–11, 1973. H. Huber.
2. Anido, G., van Kampen, E. J., Rosalki, S. B. & Rubin. M. (eds.) (1975) Quality Control in Clinical Chemistry. Transactions of the 6 th International Symposium, Geneva, April 23–25, 1975. W. de Gruyter, Berlin–New York.
3. Boutwell, Jr., J. H. (1975) A Modest Proposal: Specifying Sensitivity, Clin. Chem. *21*, 1542.
4. Bowers, Jr., G. N., Bergmeyer, H. U. & Moss, D. W. (1975) Provisional Recommendation (1974) on IFCC Methods for the Measurement of Catalytic Concentration of Enzymes, Clin. Chim. Acta *61*, F11–F24; Z. Klin. Chem. Klin. Biochem. *13*, 471–478; Clin. Chem. *22*, 384–391 (1976).
5. Broughton, P. M. G. & Annan, W. (1971) Measurement of Analytical Precision in Clinical Chemistry, Clin. Chim. Acta *32*, 433–441.
6. Broughton, P. M. G. & Raine, D. N. (1969) Quality Control in Clinical Biochemistry, Ann. Clin. Biochem. *6*, 89–148.
7. Büttner, J., Borth, R., Boutwell, J. H. & Broughton, P. M. G. (1975) Provisional Recommendations on Quality Control in Clinical Chemistry. Part 1. General Principles and Terminology, Clin. Chim. Acta *63*, F25–F38, following p. 108; Z. Klin. Chem. Klin. Biochem. *13*, 523–531; Clin. Chem. *22*, 532–540 (1976).
8. Büttner, J., Borth, R., Boutwell, J. H., Broughton, P. M. G. & Bowyer, R. C. (1976) Provisional Recommendations on Quality Control in Clinical Chemistry. Part 2. Assessment of Analytical Methods for Routine Use, Clin. Chim. Acta *69*, F1–F17, following p. 152; J. Clin. Chem. Clin. Biochem. *14* 265–275; Clin. Chem. *22*, 1922–1932. – (1979), Approved Recommendation, Clin. Chim. Acta *98*, 145F–162F; J. Clin. Chem. Clin. Biochem. *18*, 78–88.
9. Büttner, J., Borth, R., Boutwell, J. H., Broughton, P. M. G. & Bowyer. R. C. (1977 a) Provisional Recommendations on Quality Control in Clinical Chemistry. Part 3. Calibration and Control Materials, Clin. Chim. Acta *75*, F11–F20, following p. 180; J. Clin. Chem. Clin. Biochem. *15*, 233–238; Clin. Chem. *23*, 1784–1789.
10. Büttner, J., Borth, R., Boutwell, J. H., Broughton, P. M. G. & Bowyer, R. C. (1977 b) Provisional Recommendations on Quality Control in Clinical Chemistry. Part 6. Quality Requirements from the Point of View of Health Care, Clin. Chim. Acta *74*, F1–F9, following p. 190; J. Clin. Chem. Clin. Biochem. *15*, 95–100; Clin. Chem. *23*, 1066–1071.

10a. Büttner, J., Borth, R., Boutwell, J. H., Broughton, P. M. G., & Bowyer, R. C. (1978) Provisional Recommendation on Quality Control in Clinical Chemistry. Part 5. External Quality Control. Clin. Chim. Acta *83*, 189F–202F; J. Clin. Chem. Clin. Biochem. *16*, 259–266; Clin. Chem. *24*, 1213–1220.

11. Büttner, H., Hansert, H. & Stamm, D. (1974), Statistical Analysis, Control and Assessment of Experimental Results. In: Methods of Enzymatic Analysis (Bergmeyer, H. U., ed.), Vol. 1, pp. 318–395, Weinheim, and New York–London.
12. Cali, J. P., Bowers, J. N & Young, D. S. (1973) A Referee Method for the Determination of Total Calcium in Serum, Clin. Chem. *19*, 1208–1213.
13. Cooper, G. R. (1969) Quality Performance in Clinical Pathology, Progr. Clin. Pathol. *3*, 1–71.
14. Cotlove, E., Harris, E. K. & Williams, G. Z. (1970) Biological and Analytic Components of Variation in Long-Term Studies of Serum Constituents in Normal Subjects. III. Physiological and Medical Implications, Clin. Chem. *16*, 1028–1032.
15. Dixon, K. & Northam, B. E. (1970) Quality Control Using the Daily Mean, Clin. Chim. Acta *30*, 453–461.
16. Eisenhart, C. (1968) Expression of the Uncertainties of Final Results, Science *160*, 1201–1204.
17. Harris, E. K., Kanofsky, P., Shaker, G. & Cotlove, E. (1970) Biological and Analytic Components of Variation in Long-Term Studies of Serum Constituents in Normal Subjects. II. Estimating Biological Components of Variation, Clin. Chem. *16*, 1022–1027.
18. Ingle, J. D. (1974) Sensitivity and Limit of Detection in Quantitative Spectrometric Methods, J. Clin. Educ. *51*, 100–105.
19. International Symposium on Statistical Quality Control in the Analytical Laboratory (1968) Z. Anal. Chem. *243*, 751–825.
20. International Union of Pure and Applied Chemistry, Analytical Chemistry Division, Comission on Spectrochemical and Other Optical Procedures for Analysis (1972) Nomenclature, Symbols, Units and their Usage in Spectrochemical Analysis. II. Terms and Symbols Related to Analytical Functions and their Figures of Merit. Information Bulletin, Appendices on Tentative Nomenclature, Symbols, Units, and Standards. Number 26
21. Lang, H., Rick, W. & Róka, L. (1973) Optimierung der Diagnostik (Merck-Symposium 1973 of the German Society of Clinical Chemistry), Springer-Verlag, Berlin
22. Meinke, W. W. (1971) Standard Reference Materials for Clinical Measurement, Anal. Chem. *43*, 28A–47A.
23. Pickup, J. F., Jackson, M. J., Price, E. M., Healy, J. R. & Brown, S. S. (1975) Use of the Reference Method for Determination of Serum Calcium in a Quality-Assurance Survey, Clin. Chem. *21*, 1416–1421.
24. Proceedings: International Conference on Standardization of Diagnostic Materials. June 5–8, 1973. U.S. Department of Health, Education and Welfare (Oct. 1974). (Obtainable from Centre for Disease Control, Atlanta, GA, U.S.A.)
25. Rappoport, A. E. (ed.) (1972) Quality Control in Clinical Chemistry. Transactions of the 5 th International Symposium, Geneva, May 12–14, 1971, H. Huber.
26. Report prepared by the Analytical Standards Committee (1965) Analyst (London) *90*, 251.
27. Stamm, D. (1974) Calibration and Quality Control Materials, Z. Klin. Chem. Klin. Biochem. *12*, 137–145.

28. Stevens, J. F. (1973) Control Materials for Clinical Biochemistry, Tech. Bull. No. 29, Ann. Clin. Biochem. *10*, 133–145.
29. Tonks, D. B. (1963) A Study of the Accuracy and Precision of Clinical Chemistry Determinations in 170 Canadian Laboratories, Clin. Chem. *9*, 217–233.
30. Westgard, J. O., Carey, R. N. & Wold, S. (1974) Criteria for Judging Precision and Accuracy in Method Development and Evaluation, Clin. Chem. *20*, 825–833.
31. Whitby, L. G., Mitchell, F. L. & Moss, D. W. (1967) Quality Control in Routine Clinical Chemistry, Adv. Clin. Chem. *10*, 65–156.
32. Williams, G. Z., Young, D. S., Stein, M. R. & Cotlove, E. (1970) Biological and Analytic Components of Variation in Long-Term Studies of Serum Constituents in Normal Subjects. I. Objectives, Subject Selection, Laboratory Procedures and Estimation of Analytic Deviation, Clin. Chem. *16*, 1016–1021.
33. Wilson, A. L. (1970) The Performance Characteristics of Analytical Methods. 1, Talanta *17*, 21–29
34. Wilson, A. L. (1970) The Performance Characteristics of Analytical Methods. 2, Talanta *17*, 31–34
35. Winkel, P., Statland, B. E., Bokelund, H. & Johnson, E. A. (1975) Correlation of Selected Serum Constituents: 1. Inter-Individual Variation and Analytical Error, Clin. Chem. *21*, 1592–1600.
36. Young, D. S., Harris, E. K. & Cotlove, E. (1971) Biological and Analytic Components of Variation in Long-Term Studies of Serum Constituents in Normal Subjects. IV. Results of a Study Designed to Eliminate Long-Term Analytic Deviations, Clin. Chem. *17*, 403–410.
37. Young, D. S. & Mears, T. W. (1968) Measurement and Standard Reference Materials in Clinical Chemistry, Clin. Chem. *14*, 929–943.

Part 2. Assessment of Analytical Methods for Routine Use

This paper is based on a Provisional Recommendation (Büttner et al. 1976), which has been revised to take account of the comments received, and has been accepted by the Council of the International Federation of Clinical Chemistry by mail ballot 1978-06-15.

Contents

1. Introduction

1.1

Most quality control (QC) procedures are concerned with monitoring the precision and accuracy of analytical methods. These procedures should be capable of detecting changes in performance, but no system of QC is likely to improve a method which is fundamentally unsound. Information about the performance of an analytical method must therefore be obtained before it is used routinely with patients' specimens. The selection of a method, and the criteria used in making this choice, are therefore of fundamental importance in QC.

1.2

The features used to describe the characteristics of an analytical method (4) are its practicability (speed, cost, technical skill requirements, dependability and laboratory safety) and reliability (specificity, accuracy, precision, and detectability). When considering the suitability of a method, the analyst requires information about all these features, and if this is not available, it must be obtained by well-defined procedures and, whenever possible, be in the form of numerical data, so that judgement can be made objectively. With this information, the analyst can select a method which will meet his required standards of performance, which will depend on the purpose for which the method is to be used.

1.3

The assessment of an analytical method is a continuing process during which information about its characteristics is gradually accumulated. Some data will have been provided by the originator of the method, and these may be checked and supplemented by other clinical chemists who use it. If the method is considered for official recommendation, it is likely to undergo further testing. At each stage, information about the characteristics of the method is produced, and if this can be obtained, assembled and presented in a standard manner, which is meaningful to all clinical chemists, unnecessary duplication of effort can be avoided.

1.4

Errors may arise either from the method or from the individual laboratory or analyst using it, and these can only be distinguished by repeating at least part of the assessment in other laboratories. Collaborative testing is therefore an important part of the assessment.

1.5

This document describes a series of tests which can be used both by the originator of a method and by others who try it out in their own laboratories. It gives detailed recommendations on the procedures to be used to obtain performance data, and for presenting this information. The recommendations apply to methods intended for routine use, and may not be applicable to definitive or reference methods.

2. Principles of assessment

2.1

Data about the performance characteristics of analytical methods must be obtained by standardized procedures, described in such a way that other analysts can repeat them, and presented in a form which will permit comparison with other methods and with stated standards of performance.

2.2

The analytical methods used in clinical chemistry differ widely and involve measurements of such diverse quantities as concentration, enzyme activity, ionic activity, and partial pressure. Some depend on complex equipment which can be evaluated independently, but details of instrument evaluation are not considered here. It should be recognized that the general procedures described in this paper may not all be applicable, or relevant, to all methods, and some may require modification before they can be applied to particular analytical methods.

2.3

Since results may become better with practice, a period of familiarization with the method is advisable before commencing the formal assessment. Unless the report of the assessment states otherwise, the reader will normally assume that the results quoted are typical of those obtainable with the method under routine conditions.

2.4

The performance characteristics of methods are often interrelated: for example, precision may be affected by speed, and accuracy by specificity and technical skill. The assessment should be designed to obtain information about factors which are inter-dependent.

2.5

In this paper, test procedures are described in detail under four headings: practicability (Section 4), precision (Section 5), accuracy and specificity (Section 6), and detectability (Section 7). Tests need not be carried out in the order described: thus data on practicability can be obtained during the assessment of accuracy and precision.

2.6

Since a complete evaluation is lengthy, it may be useful to make a preliminary assessment of the method using only some of the tests described, and then, using all available information (including that provided by the originator of the method), decide whether to proceed further. At any time during the assessment, it may be decided that the method is impractical or that the results are unsatisfactory. The following options must then be considered:

2.6.1. Should the method be investigated further and, after identification of the cause of failure, be suitably modified? If this is done, the amended method will have to be fully described and tested.

2.6.2. Should the method be rejected? Before this is done, and the assessment abandoned, careful consideration should be given to the possible acceptability of the method in different circumstances. For example, a manual method may be rejected by a laboratory because the results are inferior to those of an automated method, whereas another laboratory without automatic equipment may consider it preferable to a worse manual method (or to no method at all). For this reason, all assessments should be continued until sufficient data are obtained to allow any laboratory to judge the acceptability of the method in local circumstances.

3. Description of the analytical method

3.1

An analytical method is defined as a set of written instructions which describe the procedure, materials and equipment which are necessary for the analyst to obtain a result. No meaningful assessment is possible unless all relevant details are fully specified.

3.2

The following information should be given:

(a) Outline of the principle, with references to the literature.

(b) Specifications of instruments and equipment, given in sufficient detail to allow the analyst to judge which commercial instruments or models can or cannot be used.

(c) List of reagents with concentrations and their tolerance limits expressed in appropriate SI units. Source, brand name and, if possible, purity should be indicated. The method used to derive the concentration of calibration standards should be stated.

(d) Specimen requirements, including volume, nature of any preservatives or anticoagulants, recommended storage conditions, and characteristics of unsuitable specimens.

(e) Full description of all steps in the analytical procedure. This should include a clear indication of critical steps and the tolerance which may be allowed to measurements (such as volume, time, temperature, wavelength, etc.).

(f) Calibration procedure and method for calculating results.

(g) Analytical range, that is, the range of concentration or other quantity in the specimen over which the method is applicable without modification. For most methods, alteration of the specified sample volume constitutes a modification. Accuracy and precision deteriorate at both ends of the range, and procedures must be described for dealing with specimens outside the range.

(h) Special safety precautions in handling specimens, preparing reagents and disposing of waste, and appropriate methods of decontamination.

3.3

Too much detail is preferable to too little. If some feature of the method is not specified in detail, the analyst is justified in assuming that it is unimportant. The assessment will be valueless if different analysts interpret the instructions in different ways or introduce their own modifications.

3.4

If the important variables of the method to be assessed (in this document called the candidate method) are not known, a preliminary examination must be made to obtain this information.

Practicability

4.1

The experience gained during the initial familiarization period, as well as in the assessment itself, will give information about the practicability and potential usefulness of the method.

4.2 Speed

Two descriptive time-related variables are relevant: the rate at which specimens can be processed, and the interval between presentation of the specimen to the analyst and the availability of the result. Three measurements are required:

(a) The time needed for analysis of one specimen when the analyst is not prepared for it.

(b) The time needed for analysis of one specimen when the analyst is prepared for it.

(c) The number of specimens which can normally be analyzed per hour, day, or week.

If speed can be varied, the effect of this on accuracy and precision should be assessed.

4.3. Cost

It is essential that the costing methods used, together with any assumptions made, are fully described so that data from different assessments can be compared. The cost of reagents and labour, capital and running cost (consumables, repairs, maintenance) of equipment, as well as general overhead should all be listed separately. The total cost per specimen or per test should be stated for the maximum rate at which specimens can be processed, for smaller work loads (e.g. 50% and 20% of the maximum), and for a single specimen.

4.4 Technical skill requirements

The initial testing of a method may involve staff more senior than that later found adequate to operate the method routinely. If precision and accuracy depend on special skills, these should be noted, and performance data should be obtained by the type or grade of staff who would use the method routinely.

4.5. Dependability

This feature includes the mechanical and electrical reliability of equipment, and the risk or frequency of breakdowns. The frequency of breakdown or method failure under routine conditions should be noted.

4.6 Safety

An assessment should be made of the electrical and mechanical safety of any equipment used; of chemical hazards such as fire and explosion; of poisonous, carcinogenic, or radioactive reagents; and of microbiological hazards arising from specimens and reagents. In addition, an assessment should be made of the hazards which may arise from accidents such as reagent spillage or equipment malfunction, and the steps necessary to deal with them should be described.

4.7 Comment

At least some of these practicability features will normally have been examined by the originator of the method, but his claims should be checked. With some aspects, such as technical skill requirements and safety, it may be difficult to obtain objective data; subjective opinions are useful provided, that they are clearly identified as such. Where opinions differ, or performance data conflict, further investigation is required. Dependability may be difficult to assess from the brief experience of one laboratory, and a full assessment may only be possible from the collected experience of several clinical chemists over longer periods. Comments from users who have abandoned the method as impracticable may be just as useful as those from others who have found it satisfactory.

5. Precision

5.1

Precision is assessed from the results of replicate analyses made according to a well-defined experimental design. The factors which contribute to imprecision were listed by Büttner et al. (Ref. 4, Section 2.2.) and during the evaluation as many as possible of these should be kept constant by, for example, using the same batch of reagents, and the same standards, instruments, and analysts, although, of course some studies must also be made of the effect of varying these factors. Precautions should be taken to ensure that accuracy does not change during the trial period. Two experimental designs are suggested here. In one, imprecision is measured from the differences between duplicate results obtained with a number of different specimens, and in the other it is derived from the results of several replicate measurements made on the same specimen. The nature of the specimen and the arrangement of the replicates are important with both designs. Preferably, replicates should not be identifiable by the analyst.

5.1.1. Ideally, a variety of patients' specimens, typical of those encountered in routine practice, should be used for the assessment. However, these are usually unstable and consequently can only be used if the replicate analyses are performed within a short period of time (e.g. within a run or a day). Control specimens are usually more stable and are easier to use when studying the effect of concentration on imprecision, but the results will only be applicable to the specimens used. In each case, precautions should be taken to ensure that specimens are stable over the period of analysis. Replicates should be arranged in random positions in a run of patients' specimens, and not in a sequence, so that carryover effects are included.

5.1.2. Two experimental procedures are described in Section 5.2 and 5.3 to illustrate these principles; modifications and combinations are possible. The results obtained by these procedures should not be compared or confused, but if both are used, more information will be obtained about the factors which affect precision.

5.2. Within-run imprecision

5.2.1. This procedure is useful for making a rapid assessment of the imprecision of a method and in comparing the accuracy of two methods. Select a number (up to 100) of patients' specimens and analyze these in duplicate by the candidate method and by the comparison method. The specimens chosen should span the analytical ranges of both methods, and note should be made of characteristics (such as the presence of drugs, lipaemia, haemolysis, jaundice, or uraemia) which might be relevant in the assessment of accuracy (see Section 6.8.4). Specimens may be analyzed in one or more runs, but duplicates must be included in the same run.

5.2.2. The differences between duplicates are used to calculate the standard deviation (S.D.) of the candidate and the component present, S.D.s are calculated from the results for (1) all specimens, (2) specimens within the lower (defined) part of the analytical range, and (3) the upper (defined) part of the range. If necessary, the range may be further subdivided. The mean and number of duplicate pairs are recorded for each range, and the differences in variation tested for statistical significance.

5.3. Between-day imprecision

5.3.1. Select a number (not less than 3) of specimens which span the analytical range of the method; the exact number will depend on the rate of change of imprecision with concentration or other quantity measured, which will be indicated by the results of the tests described in Section 5.2. At least one specimen should have a low value, one a high value, and one an intermediate value. If specimens are liquid, about 20 replicate samples from each should be separately stored deep-frozen and one thawed each day for analysis. If lyophilized serum is used, the same recommended reconstitution procedure should be used each time. Alternatively, a sufficient number of vials should be reconstituted, the contents pooled and then treated as a liquid specimen.

5.3.2. One sample of each specimen is analyzed, in a random position in a run of patients' specimens, once per day for, say, 20 consecutive working days. If the analysis is not performed every day, the test should be made for the 20 consecutive days on which the analysis is run. The results should be plotted to check for changes of accuracy with time, and, if necessary, further replicate analyses are made during a period of constant accuracy.

5.3.3. The S.D. of the 20 results is calculated after excluding identifiable errors. The results are expressed as the between-day S.D., and the mean value for each specimen is quoted. The number of replicates must be stated; in most cases, 20 is an adequate though arbitrary number.

5.4

Imprecision can be influenced by interfering compounds and, for enzyme assays, by the presence of isoenzymes in variable proportions. Any evidence about these and other effects should be given. Before a method is recommended for acceptance, tests should be repeated independently in other laboratories. If the results agree, they may be taken as typical of the method; if they do not, further investigation is required.

6. Accuracy and specificity

6.1

Basically the assessment is concerned with only one question: how accurate is the method? This may be answered by one set of tests, whereas different tests may be required if attempts are made to identify the sources of inaccuracy. Selection of assessment procedures may therefore depend on circumstances as well as the type of method. Simple tests for inaccuracy (e. g. linearity and recovery checks) should always be made before more complex procedures are undertaken. It must be emphasized, however, that they will only detect inaccuracy arising from certain causes. If these tests prove satisfactory, these sources of inaccuracy are not present, but the method may still be inaccurate from other sources.

6.2. Calibration standards

The accuracy of a set of results will be affected by errors of calibration. When comparing two methods, it is preferable to use the same set of calibration standards for both. If this is not possible, both sets of standards should be analyzed by both methods to obtain a measure of any difference between them. If a method is calibrated with an aqueous standard, there is little purpose in analyzing other aqueous solutions of known concentration to test accuracy. If some other calibration procedure is used, a pure material or solution should be analyzed to test the accuracy of this calibration. In some methods it is impossible to put a standard through all stages of the analytical method, and this, of itself, may be a source of inaccuracy.

6.3. Linearity

Some methods do not employ a full range of standards to obtain a calibration curve, but depend instead on single-point or two-point calibration. Any non-linearity will produce inaccurate results between the two fixed points. Linearity should always be tested, and not assumed. One way of checking this is to analyze several dilutions of a specimen, or graded mixtures of two specimens, one with a high value and the other with a low value. Since enzymes may be inactivated by dilution in solvents, linearity of enzyme activity should be tested with graded mixtures of two specimens and by measuring activity-time curves for specimens with different enzyme activities.

6.4. Recovery

Recovery experiments depend on the analysis of two samples which are identical except that a known amount of pure material is added to one sample (Ref. 4, Section 3.3.2.). They are particularly useful in testing the various stages of an analytical method and for assessing losses in, for example, chromatography or solvent extraction. In designing recovery experiments, the following points should be considered.

6.4.1. It may be difficult to dissolve the pure material in the sample. If it is added in the form of a solution, the samples will no longer be identical unless the same volume of solvent is added to both.

6.4.2. The pure component added may be in a different form from that naturally present (e.g. protein-bound or conjugated). It may then be better to limit the recovery experiment to the appropriate part of the method, by, for example, making the addition after hydrolysis of the natural conjugated component.

6.4.3. The two samples will have different concentrations, and both must be within the analytical range of the method. The precision of both estimates will affect the confidence limi ts of the recovery. Normally, several recovery experiments should be made by adding varying amounts of pure material to each specimen.

6.4.4. The specificity of a method depends on the type of specimen analyzed. Consequently, recovery experiments should be made with as many specimens as possible, in order to include some containing potentially interfering substances.

6.4.5. If specimens can be obtained or prepared which are free of the analyte, the sample without addition will provide a true sample blank value. This is particularly useful with methods for determining exogenous components such as drugs.

6.5. Blanks

Many methods require a blank analysis in order to correct for non-specificity. These are of two types. A reagent blank is used to correct for the contribution of the reagents to the reading of sample solutions. A sample blank is used to compensate for the reading of sample solutions. A sample blank is used to compensate for the reading given by substances other than the analyte, which are present in the sample. In some situations, an analyte-free sample can be used (see Section 7.7), but usually, sample blanks are prepared by omitting an essential reagent or step in the analysis, and these will include at least part of the reagent blank. In most cases, the magnitude of the sample blank varies between samples and must be measured for each one. However, if this variation is shown to be negligible, a single specimen (for example, one from which the analyte is absent or has been removed) can be analyzed with each run or as required. The validity of the blank procedure given in the method should always be tested. Often, errors revealed by the other assessment procedures can be traced to inadequate blank compensation.

6.6. Specificity

Some methods give inaccurate and falsely high results (Ref. 4, Section 3.2.2.1) because sample components other than the analyte contribute to the reading. The limitations of specificity in a method should be given in the description of the method, but may still require confirmation or more detailed assessment. In principle, this assessment is open-ended, because any food or drug must be suspected of contributing to the reading unless proved innocent.

6.7. Interference

This is the effect of a component, which does not by itself produce a reading, on the accuracy of measurement of another component. It results in either low values for the analyte (inhibition) or high values (enhancement). The magnitude of the effect depends on the concentration or amount of interfering material, but not necessarily in a directly proportional manner.

6.7.1. Non-specificity and interference only affect accuracy when an interfering component is present in the sample. Assessment of accuracy by analysis of pure solutions is therefore of little value.

6.7.2. Few of the analytical methods used in clinical chemistry are completely specific for one analyte and free from interference by all possible components of patients' specimens. Testing procedures should be chosen to show whether the method is sufficiently specific and free from interference to give accurate results with the specimens which are normally encountered. During the assessment it may be found that some specimens give discordant results when analyzed by different methods, and further investigations should be made to identify the component responsible. An understanding of the principle of the candidate method will often suggest possible interfering compounds. If, for example, it is suspected that bilirubin interferes with a method, it should be added in varying amounts to samples of a specimen, which are then analyzed. Specificity can be tested by adding bilirubin to a blank solution which does not contain the analyte. It may be important to distinguish between non-specificity and interference in order to overcome the inaccuracy. For example, some flame emission methods for the determination of potassium give inaccurately high results in the presence of large concentrations of sodium. If the method mistakes sodium for potassium, it is non-specific, due for example to the use of a wide-band filter. If the method will only detect potassium, and not sodium or any other atom, it is non-specific, due for example to the use of a wide-band filter. If the method will only detect potassium, and not sodium or any other atom, it is specific but inaccurate owing to interference (enhancement) by sodium. The cause of this may be depression of the ionisation of potassium atoms by sodium ions in the flame, which can be resolved by adding sodium to all potassium standards.

6.8. Comparison of methods

Accuracy may be assessed by comparing the values obtained by the candidate method with those found with a comparison method of known accuracy. All analyses should be made in replicate using specimens with concentrations spanning the analytical range. The following sections consider the problems of (1) selecting the specimens, (2) selecting the comparison method, and (3) evaluating the results obtained by the two methods.

6.8.1. Selection of specimens. Specimens must be stable during the period of the analyses by the two methods, and should be representative of those which a laboratory would expect to use with the candidate method.

6.8.1.1. Commercial control specimens with stated values usually have a defined stability and can be obtained in bulk, so that repeated analyses can be made over a period of time. They have three major limitations:

(a) The stated values may be inaccurate or given in units different (e.g. enzymes) from those used in the candidate method.

(b) Non-human components (e.g. enzymes) in the control specimen may behave differently from those in patients' specimens.

(c) Foreign material (such as preservatives) added during manufacture, and artefacts formed by processing, may interfere. Enzymes may be altered so that special treatment is needed to reactivate them. Analysis of a range of control materials can be useful in the rapid assessment of accuracy. If specimens from different sources all give results which differ consistently from their stated values, the method is unlikely to be accurate. In practice, however, this procedure often shows only that stated values differ in accuracy. If the stated values are obtained by a reference method, they will test the accuracy of the method with these specimens, but not necessarily with patients' specimens.

6.8.1.2. Analysis of *patients' specimens* is essential in the assessment of accuracy. Specimens should be selected as described in Section 5.2.1 and analyzed in replicate by both methods. Any specimens giving discordant results are re-examined, and further tests are made to identify the interfering materials. Any proven interference is recorded, and these results, if few in number, are removed before proceeding with statistical comparison of the data.

6.8.2. Classification of analytical methods in terms of accuracy

6.8.2.1. Accuracy is defined as the agreement between the mean estimate of a quantity and its true value. The accuracy of methods used in clinical chemistry can only be properly assessed with biological specimens for which the method is designed, and, since the true values of these are not known, estimates must be made by another method. Assessment of accuracy thus depends on a comparison of two values of the same specimen, one obtained by the candidate method, and the other by a comparison method which ideally should give true values. The accuracy of the comparison method is therefore of fundamental importance.

6.8.2.2. Analytical methods may be classified into 4 types, according to their degree of accuracy (Ref. 4, Section 5).

(a) A *definitive method* is one which, after exhaustive investigation, is found to have no known source of inaccuracy or ambiguity. The result it gives is termed the *definitive value* and is the best known approximation to the true value. An example is the isotope dilution–mass spectrometry method for determining serum calcium concentration (8).

(b) A *reference method* is one which after exhaustive investigation has been shown to have negligible inaccuracy in comparison with its imprecision. An example is the determination of serum calcium concentration by the atomic absorption spectroscopy method of *Cali* et al. (8). Since there may be alternative reference methods for the same analyte, the reference to the publication should always be given. The *reference method value* (which may be stated or certified) is that derived from a set of results obtained by a reference method.

(c) A *method of known bias* is one in which the inaccuracy has been established (for example, by comparison with a reference method); it may be zero. The result it gives with a standard or control sample is termed an *assigned value* (which may be stated or certified), and its confidence limits should be given. The amount of bias may depend on concentration or on the presence of interfering materials. If results are corrected for the bias, this must be stated and the derivation of the correction explained.

(d) A *method with unknown bias* is one of unknown accuracy and the result it gives is an *assigned value* (stated or certified).

6.8.3. Selection of comparison method

6.8.3.1. Definitive methods must be used in the assessment of reference methods, but will not be considered further here.

6.8.3.2. A *reference method* could be used as a comparison method, but it may be difficult to use and available in only a few centres. It would be possible for another laboratory to use a reference method to fix values to a number of specimens which could then be analyzed by the candidate method. At best, a wide range of patients' specimens should be used, but since reference methods are often laborious, this may be expensive and time consuming. In practice it is more likely that the stated or certified values of control specimens will be used, despite their inherent limitations.

6.8.3.3. Methods with known bias should only be considered for comparison purposes when reference methods are not available. The inaccuracy will be known for certain types of specimens, but may not be constant for all specimens. Consequently, the comparison of values may only be valid for a limited range of specimens. In this situation, it may be advisable to make the comparison against two or more methods of known bias (see Section 6.9)

6.8.4. Procedure. Analyze a number (e.g. about 100) of patients' specimens in replicate by the candidate method and the comparison method, as described in Section 5.2.1. The results are then examined by statistical procedures described below.

6.8.5. Evaluation of results

6.8.5.1. While specificity, precision, and detectability pose no special statistical problems, tests for the comparison of inaccuracy of two methods are more difficult. These tests have two objectives, which are related but not identical.

(a) One objective is to investigate the predictive power of the candidate method under the assumption that it will be acceptable if, and only if, the results provide a reasonably close approximation to those of the comparison method, which is assumed to be of known reliability. The ideal relation would be a straight line through the origin with a slope of 45 ° and a residual variation similar to replicate variation, indicating essential identity of results.

(b) The other objective is to obtain an estimate and confidence limits of the mean ratio of results given by the candidate and comparison methods, and to test whether this ratio differs from unity. However, this objective makes sense only if the relation is linear and the line passes through the origin.

6.8.5.2. It should be emphasized that coefficients of correlation and ordinary regression equations (whether of "y on x" or "x on y") are of very little use with regard to these two objectives. Scatter diagrams may be convenient as graphical summaries of comparison data, but significant correlation will frequently be a foregone conclusion anyway, and correlation may be highly significant (in the sense of having a very low probability of chance occurrence) while being much too weak for prediction within limits narrow enough to approach even remotely the precision required in measurement. Moreover, the assumptions usually postulated or implied in the ordinary regression model (distinction between dependent and independent variable; values of independent variable controlled with negligible error) are not met when two analytical methods are being compared. In this situation, the problem is the determination of the functional relationship between two variables when both are subject to error, and one of the statistical methods available to handle this problem should be used.

6.8.5.3. When evaluating comparison data on accuracy, it is recommended that the following sequence of steps be followed, in order not to overlook any of the basically different structures which the data may exhibit.

(a) Plot the data pairs as points on a scatter diagram, using the abscissa (x) for the comparison method and the ordinate (y) for the candidate method. The diagram will provide a useful first impression of the type of relationship between the two methods. If the relationship is clearly curved rather than linear, further analysis in unlikely to lead to adoption of the candidate method on the basis of the comparison data alone, although there are, of course, statistical methods for dealing with non-linear curves.

(b) Plot the absolute difference, $|y-x|$, against the mean, $(x+y)/2$, of each pair. Take logarithms of the original data, and again plot the differences, $|\log y - \log x|$, against the means, $(\log y + \log x)/2$. Draw freehand curves through the two scatter diagrams. If the ordinary differences increase with the mean values, while the logarithmic differences are reasonably constant over the the whole range, all calculations should be done with the logarithms, in order to equalize the variance. There are, of course, standard statistical methods to obtain and test homogeneity of variance with more rigour, but this is rarely necessary. Due to the law of mass action, experimental variation in the results of chemical analysis will make distributions symmetrical over the range considered and variances reasonably constant. Some care is needed when results from an analysis of logarithms are reconverted and reported in terms of the original quantities.

(c) Using appropriate statistical methods, calculate the linear functional relation between x and y (or $\log x$ and $\log y$), and test whether the intercept differs statistically from zero. If it does, use professional judgement to decide whether it is small enough to be disregarded for all practical purposes. This procedure will either permit or invalidate the conclusion that the line goes through the origin.

(d) If the line passes through the origin, use the logarithmic differences to calculate the mean and the tolerance range of the ratio y/x. This will show whether the variation in the ratio y/x and the mean deviation from unity are both small enough to be negligible for analytical purposes. If the ratio does differ more than negligibly from unity, the candidate method is not acceptable, until the reasons for this proportional bias are found and eliminated.

(e) If the line does not pass through the origin, the candidate method has a constant bias with respect to the comparison method. The bias may or may not be acceptable. If it is, the equation obtained in (c) should be used to calculate predicted values of x from observed values of y, and the confidence limits of such estimates or predictions. Professional judgement must be used to decide whether these limits are narrow enough for analytical purposes, i.e. whether the variation obtained in the comparison is reasonably close to the analytical imprecision of either method.

6.9. Collaborative testing of methods

6.9.1. The performance of two methods within one laboratory may be influenced by local conditions, and another laboratory using the same pair of methods may reach different conclusions. It is therefore important that tests be made in several laboratories, and this is particularly useful in assessing the importance of reagents, standards, equipment, and technical skill, and the adequacy of the written instructions. It is also invaluable in giving more information on safety, costs, and dependability.

6.9.2. At present, definitive and reference methods are not available for the assay of many components, and collaborative testing is a useful method of deriving consensus values for specimens, to be used in assessing accuracy. Several external QC surveys have shown that laboratories using the same method to analyze the same specimen may obtain different results. For many components, the scatter of results is large. However, if a sufficient number of these laboratories later re-analyze the same specimen, the values (after exclusion of outliers) are often consistent. The mean is a consensus value. If a sufficient number of independent methods all give the same (consensus) value on a specimen, either this value must be the best estimate of the true value or all the methods must be equally inaccurate. If the methods are independent, the former alternative is the most likely, and this hypothesis could be tested for assays where definitive or reference methods are available (11). In other cases, the consensus value for each set of methods will require close scrutiny before it is accepted.

6.9.3. This procedure could be used to assess the accuracy of a candidate method for a component which is assayed by many laboratories using independent methods. A variety of patients' specimens could be used and, since each method would be used simultaneously in a number of laboratories, the results would be representative of the method and not of single laboratories.

7. Detectability

7.1

Intuitively, a "sensitive" analytical method is one which has the ability to detect small amounts of analyte. In this document, the term *detectability* is used for this performance characteristic, in order to avoid the confusion which has arisen from the unfortunate fact that the term *sensitivity* is used in analytical chemistry for several quite different concepts, depending on whether interest is focussed on the final result obtained by a complete method, or on measuring instrument performance in terms of readings and other calibration variables. Clearly the two are intimately related: instrument performance is part of, and a major factor in, whole-method performance.

7.2

For example, the slope of the calibration curve has been called "sensitivity" to express the concept that, all other factors (especially variation) being equal, a greater (steeper) slope permits the detection of smaller differences in analyte concentration for a given increment in readings (signal). The ratio slope/S.D. (where S.D. is the standard deviation of replicate readings) has also been called "sensitivity" in recognition of the fact that, in detecting small variation can be just as effective as, or even more effective than, a steep slope combined with large variation. Both these "sensitivity" concepts and variants of them are limited to one mode of measurement at a time, and the units are ambiguous. Moreover, it can easily be shown that these concepts are closely similar in factual content to the performance characteristic which, without ambiguity or disagreement, is commonly referred to as precision. Thus, the ratio S.D./slope has been used for many years as the index of precision in the statistics of bioassay, because, for the logarithmic dose scale customary in that field, this ratio is the S.D. of individual log doses, and is, therefore, a useful precision parameter which is invariant with respect to definition of response and unitage of analyte.

7.3

The use of a second term (sensitivity) for the concept adequately named *precision* should be discouraged and, ultimately, discontinued. Nevertheless, while not endorsing that usage, we recommend the term *detectability* for the concept previously defined (4) as one of the four reliability criteria, namely "the ability of an analytical method to detect small quantities of measured component".

7.4

Detectability, like precision and accuracy, has no numerical value. The corresponding measured quantity is the detection limit, which has been defined (4) as "the smallest single result which, with a stated probability, can be distinguished from a suitable blank".

7.5

It is clear from this definition that the detection limit results from the interaction of several factors which should all be considered, and in fact always are considered implicity if not explicitly, when an analytical method is set up for routine use. The calibration process yields the working range of analyte concentrations which give meaningful readings with the standard and the equipment used. At both ends of the working range, the calibration curve deteriorates progressively: at the lower end, because standard readings overlap with reagent blank readings; at the upper end, because imprecision increases unacceptably or the curve flattens. Sample volume is chosen so that most expected sample readings will fall within the working range of the calibration curve, taking into account expected sample values, average recovery (if relevant), and the fact that it is bad practice to operate an analytical method close to either limit of the working range. The working range of calibration standards and the sample volume can be adequately determined in this way without mathematical rigour or statistical testing; once determined, they become part of the method.

7.6

It is then necessary to determine the detection limit which will serve three purposes:

(a) To characterize the detectability of the method as compared with other methods for the same analyte;

(b) to avoid frequent misuse of the method in attempts to measure values near or lower than the detection limit; and

(c) to recognize those results which should be reported as "lower than . . . (detection limit)" rather than as numerical values. Here the detection limit is an insurmountable barrier, whereas values above the working range can usually be determined by using diluted specimens or smaller sample volumes.

7.7

The determination of detection limits requires the following steps:

(a) Select suitable blank samples. Ideally, these differ from other samples only in that they are known not to contain the analyte. Unless the analyte is a drug not normally present in body fluids, the ideal requirement can only be approximated; for example, by using pretreated samples (i.e. after removal or destruction of the analyte); or samples taken from patients devoid of the tissue secreting the analyte; or by omitting a crucial reagent or step in the analysis of ordinary samples.

(b) Determine the mean and S.D. of apparent blank-sample results.

(c) Determine the S.D. of replicate results for samples with values near the blank-sample range, i.e. lower-range imprecision.

(d) Select a suitable one-sided statistical test, and an acceptable probability of chance occurrence as the significance level.

(e) Calculate the limit at which a single result would become significantly greater than the mean apparent blank-sample result. This is the detection limit.

7.8

Assuming equality of blank-sample and low sample S.D. values and 20 degrees of freedom, using a one-sided t test at $P = 0.01$, and stipulating that numerical accuracy to more than two significant digits is rarely warranted, then the detection limit would be approximately equal to *mean blank result* + 2.6 S.D. More precise or rigorous calculations are rarely necessary if it is remembered that the detection limit should be used as a warning sign (KEEP OFF) rather than a justification for continually using a method at the lowest level it may be able to measure.

8. Format of the assessment report

A report on the assessment of an analytical method should include the following information in sufficient detail so that clinical chemists can interpret it unambiguously, repeat the experiments if they so wish, and compare the data with desirable standards of performance.

8.1. Complete description of the method including the features outlined in Section 3.

8.2. Assessment of practicability, including speed, cost, technical skill requirements, dependability, and laboratory safety, based on the experience of laboratory staff who have used it (Section 4).

8.3. Measurement of within-run and between-day imprecision (Section 5) with different specimens and concentrations, using a clearly defined protonol.

8.4. Assessment of accuracy, including a comparison with values obtained on patients' specimens by an established method of known accuracy (Section 6).

8.5. Statement of substances found to interfere with the analysis (Section 6.7.2).

8.6. Comparison of results with those obtained in other laboratories (Section 6.9).

8.7. Assessment of detectability (Section 7).

8.8. Reference values ("normal range") of the candidate method should be determined. If these differ significantly from those which are already well validated, further investigation is required.

9. References

1. Barnett, R. N. & Youden, W. J. (1970) A Revised Scheme for the Comparison of Quantitative Methods, Am. J. Clin. Pathol. *54*, 454–462.
2. Boutwell, J. H. & Mather, A. (1974) Diagnostic Methods, Proceedings of the International Conference on Standardization of Diagnostic Materials, 1973, pp. 27–41, U.S. Department of Health, Education and Welfare (Oct. 1974).
3. Broughton, P. M. G., Gowenlock, A. H., McCormack, J. J. & Neill, D. W. (1974) A Revised Scheme for the Evaluation of Automatic Instruments for Use in Clinical Chemistry, Ann. Clin. Biochem. *11*, 207–218.
4. Büttner, J., Borth, R., Boutwell, J. H. & Broughton, P. M. G. (1975), Provisional Recommendation on Quality Control in Clinical Chemistry, Part 1. General Principles and Terminology, Clin. Chim. Acta *63*, F25–F38, following p. 108; Z. Klin. Chem. Klin. Biochem. *13*, 523–531; Clin. Chem. *22*, 532–540 (1976). – (1979) Approved Recommendation, Clin. Chim. Acta *98*, 129F–143F; J. Clin. Chem. Clin. Biochem. *18*, 69–77 (1980).
5. Büttner, J., Borth, R., Boutwell, J. H., Broughton, P. M. G. & Bowyer, R. C. (1976), Provisional Recommendation on Quality Control in Clinical Chemistry, Part 2. Assessment of Analytical Methods for Routine Use, Clin. Chim. Acta *69*, F1–F17, following p. 152; J. Clin. Chem. Clin. Biochem. *14*, 265–275; Clin. Chem. *22*, 1922–1932.
6. Büttner. J., Borth, R., Boutwell, J. H., Broughton, P. M. G. & Bowyer, R. C. (1977a) Provisional Recommendation on Quality Control in Clinical Chemistry, Part 3. Calibration and Control Materials, Clin. Chim. Acta *75*, F11–F20, following p. 180; J. Clin. Chem. Clin. Biochem. *15*, 233–238; Clin. Chem. *23*, 1784–1789.

6a. Büttner, J., Borth, R., Boutwell, J. H., Broughton, P. M. G., & Bowyer, R. C. (1978). Provisional Recommendation on Quality Control in Clinical Chemistry. Part 5. External Quality Control. Clin. Chim. Acta *83*, 189F–202F; J. Clin. Chem. Clin. Biochem. *16*, 259–266; Clin. Chem. *24*, 1213–1220.

7. Büttner, J., Borth, R., Boutwell, J. H., Broughton, P. M. G. & Bowyer, R. C. (1977b) Provisional Recommendation on Quality Control in Clinical Chemistry, Part 6. Quality Requirements from the Point of View of Health Care, Clin. Chim. Acta *74*, F1–F9, following p. 190; J. Clin. Chem. Clin. Biochem. *15*, 95–100; Clin. Chem. *23*, 1066–1071.
8. Cali, J. P., Bowers, G. N. & Young, D. S. (1973), A Referee Method for the Determination of Total Calcium in Serum, Clin. Chem. *19*, 1208–1213.
9. Ingle, D. (1974), Sensitivity and Limit of Detection in Quantitative Spectrometric Methods, J. Chem. Educ. *51*, 100–105.
10. Pickup, J. F., Jackson, M. J., Price, E. M. & Brown, S. S. (1974) Assessment of the Reference Method for Determination of Total Calcium in Serum, Clin. Chem. *20*, 1324–1330.
11. Pickup, J. F., Jackson, M. J., Price, E. M., Healy, M. J. R. & Brown, S. S. (1975), Use of the Reference Method for Determination of Serum Calcium in a Quality-Assurance Survey, Clin. Chem. *21*, 1416–1421.
12. Stamm, D. (1974) Calibration and Control Materials. Z. Klin. Chem. Klin. Biochem. *12*, 137–145.
13. Vickelsöe, J., Bechgaard, E. & Magid, E. (1974) A Procedure for the Evaluation of Precision and Accuracy of Analytical Methods, Scand. J. Clin. Lab. Invest. *34*, 149–152.
14. Westgard, J. O., Carey, R. N. & Wold, S. (1974) Criteria for Judging Precision and Accuracy in Method Development and Evaluation, Clin. Chem. *20*, 825–833.
15. Westgard, J. O. & Hunt, M. R. (1973) Use and Interpretation of Common Statistical Tests in Method-Comparison Studies, Clin. Chem. *19*, 49–57.
16. Whitehead, T. P., Browning, D. M. & Gregory, A. (1973) A Comparative Survey of the Results of Analyses of Blood Serum in Clinical Chemistry Laboratories in the United Kingdom J. Clin. Pathol. *26*, 435–445.

Part 3. Calibration and Control Materials

This paper is based on a Provisional Recommendation (*Büttner* et al. 1977), which has been revised to take account of the comments received, and has been accepted by the Council of the International Federation of Clinical Chemistry by mail ballot in December 1979.

Contents

1. Introduction

1.1

Calibration standards are used in an analytical process to assign values to specimens by relating the readings or analytical responses obtained to the concentration or other quantities present.

1.2

Control materials are used to prepare specimens used in the assessment of the reliability of an analytical process, particularly *precision* and *accuracy*, and as one means of monitoring analytical performance. By appropriate managerial decisions, the information developed by the use of control specimens can be used to assure that performance requirements are met.

1.3

The term "reference material" is sometimes used to denote certain types of standard and control. This term has been defined as a well characterized material or device, usually produced in quantity and generally available, used to calibrate and/or assess a measurement system in whole or in part (*Puttock,* quoted by *Cali* 1975). This term is not recommended here because of possible confusion with the materials used to derive reference values ("normal range"), and because standards and controls have different functions and often different specifications.

1.4

The specifications of adequate calibration and control materials depend on their intended use. Both must be appropriate to the analytical method used and to the specimens analyzed. A careful consideration of the nature of the desired information is necessary to determine the required characteristics.

1.4.1. With calibration standards, the stated value must be accurate to the highest degree attainable. Any error or uncertainty in this value will be reflected in the results of patients' specimens. The method of preparing the standard, and of obtaining the stated value, must be described in such a way that the standard can be reproduced exactly.

1.4.2. It is essential that control materials and specimens are homogeneous and stable (see Sections 4.2 and 4.3). When precision alone is assessed, the value of the control specimen need not be known exactly, but for assessment of accuracy, specimens with known values are required. The degree to which these values correspond to true values will determine the validity of this assessment. Control specimens can be used to monitor changes in accuracy, even though no value has been assigned to the specimen.

1.5

Various devices and control procedures can be applied to parts of the analytical process, e. g. as instrumental checks. The recommendations given here do not include this type of quality control, but cover those materials which are analyzed as if they were patients' specimens.

2. Composition

2.1

Calibration standards and control specimens are prepared using pure substances (solutes and solvents) of known composition (such as pure water); or partially purified substances (such as enzymes and proteins); or natural products (such as human or bovine serum). Commonly, these materials are combined or treated to obtain appropriate concentrations or other quantities (see Section 5).

2.2

An analytical method involves three types of material whose composition may differ: the calibration standard, the control specimen and the patients' specimen. Differences in the chemical or physical characteristics of these materials can affect the validity of results (matrix effects).

2.3

The matrix is the milieu in which the analyte exists. For example, a pure solution of sodium chloride in water can be regarded as consisting of an analyte (Na^+) in a matrix of Cl^- and water. An ideal analytical process is insensitive to matrix effects, and as a result, the constraints on standards and control materials would be few, but with biological materials, analyzed by routine methods, matrix effects are not uncommon and the mechanisms often complex.

2.3.1. Chemical matrix effects arise from non-specificity or interference with the analytical method (2). Examples are: the effect of phosphate in decreasing the flame emission of calcium; the effect of sodium in enhancing the flame emission of potassium; the effect of protein in some bilirubin procedures; and the effect of non-glucose reducing substances in serum on the measurement of glucose by copper reduction methods. In each case, differences in the matrix of standards, controls, and specimens can affect the results.

2.3.2. Physical matrix effects are often difficult to recognize. Viscosity and surface tension can affect the rate of nebulization of solutions into a flame spectrometer. In atomic absorption spectrometry, lead in pure aqueous solution is atomized at a different rate from that in whole blood, so that standards and controls must be prepared by adding inorganic lead to a matrix of blood. Physical differences may account for differences in analytical precision found when patient specimens and aqueous solutions of similar analyte concentrations are analyzed.

2.3.3. It follows that standards and control specimens, of whatever type or combination, are useful only if demonstrably applicable to the analytical method used. Standards, controls, patients' specimens, and methods are therefore interdependent, and in the selection of one, each of the others needs to be considered.

2.4

The composition of standards and control specimens should be known as completely as possible. The source and nature of specimens and starting or base materials should be specified, and any physical or chemical modification to them be made known (see Sections 4 and 5). All additives, including preservatives, should be specified whether or not it is known or intended that they affect the analytical result.

2.5

The concentration of analyte in standards and control specimens must be controllable so that preparations can be obtained which span the entire analytical range, which may go down to zero. This is not difficult to accomplish with pure solutes and solvents, but is more difficult with preparations based upon biological materials. Attempts to deplete serum or urine of one or more constituents may affect the composition in such a way as to change the matrix effects.

2.6

Treatments such as lyophilization may introduce artefacts. The composition of the material may change with time. Storage at low temperatures or in a lyophilized condition markedly increases the stability of most constituents. A notable exception is human serum lipoprotein.

3. Classification

Both standards and control materials may be classified on the basis of (a) composition, (b) the method used to assign values, and (c) the intended use of the material.

3.1

The concentration of pure solutions is a function of the purity of analyte source and solvent, and the preparation procedure. Analysis does not enter into the designation of the value. When conditions of analysis permit, this type of standard (a primary standard) is the most direct and unambiguous means of calibration which must in any case be used with reference methods. Such solutions have the advantage of replicability, so that the stability of other materials can be established without ambiguity by direct comparison. Their use allows the investigation of interferences less ambiguously than does the use of other materials.

3.2

With many routine methods, the nature of biological samples and matrix effects, as well as procedural technological constraints, may make the use of pure standards impossible or impractical. In these instances, standards and/or control specimens are used which simulate, in one or more respects, the physical and chemical characteristics and properties of patients' specimens. Two types of material are possible.

3.2.1. Pure analyte is dissolved or dispersed in a matrix chosen to simulate, in one or more respects, patients' specimens, but whose analyte concentration is shown to be either zero or insignificant. The matrix may be undefined, such as serum or urine, or it may be a loosely defined material such as serum albumin or a gelatin solution. The utility of this type of material rests upon the means by which the analyte concentration of the base material is shown to be zero, as well as the degree to which the material is appropriate to the analytical process. The concentration of analyte is thus dependent on its purity, the method of preparation, and on the accuracy of the analysis of the analyte-free base material.

3.2.2. It is not always possible for the type of material discussed in Section 3.2.1 to provide all the properties of patients' specimens. In these cases, biological materials are used as secondary standards or control specimens. The base material, such as animal serum or tissue extracts, may be modified by depletion (dialysis, selective membrane filtration, etc.) or by additions to adjust the analyte concentrations. Other treatments or additions may be used to increase stability or to remove or simulate interferences. These materials must be analyzed to determine their values.

3.3

Calibration standards must be appropriate to the analytical process and to the specimens used. The concentration (or other) value assigned to the material must be as accurate as possible, and the uncertainty associated with the assigned value must be small compared to the analytical imprecision.

3.4

For precision control, the specimens should simulate patients' specimens as closely as possible. For serum and plasma this will include some lipaemia. Between-specimen variability must be small compared to analytical variability, and specimens should have maximum practicable stability. Control specimens of different concentrations should be provided, including some at the upper limit of the analytical range for undiluted samples.

3.5

Specimens for accuracy control may be of two types. One is similar or identical to that used for precision control, but with an assigned value of defined accuracy, either relative to the "true value" or to a defined reference method. The other type is pure analyte in defined solvent. This type of accuracy control is useful for detecting accuracy changes not easily detected by the use of biological specimens, and because it is more easily replicable. It is also useful in checking accuracy when a biological material is used as a standard.

3.6

It is worth repeating that it is essential for users of these materials to maintain an operational distinction between their use for control purposes and for calibration.

4. Specifications

4.1

The specifications for standards and control materials depend upon their application, and compatibility with the analytical process and specimens is essential.

4.2

The user of lyophilized material must recognize the need for extreme care in the quality and the quantity of reconstituting fluid as well as in the procedural details. Errors can arise from variations in the amount of liquid used to reconstitute lyophilized materials and from variations in the amount of material in each vial. These errors can be avoided by reconstituting several vials, pooling the contents, and storing aliquots under conditions which ensure their stability.

4.2.1. Specimen differences due to lack of homogeneity and (for lyophilized materials) to variations in filling and reconstitution together should not exceed ± 0,5% of the mean value (in 95% or more of the vials), corresponding to a coefficient of variations (C. V.) of 0,25%. When specimen variation (C. V.) is within these limits, there will be no significant effect on the apparent analytical precision of most methods.

4.2.2. Many routine analytical methods are not sufficiently precise to detect variability in control specimens. A useful check on the between-vial variability of lyophilized materials is to weigh the contents of, say, 10–20 vials of the same dry material, and calculate the C. V. The result will be a measure of variation in the combined amount of the solid material, and of residual moisture present.

4.3

The stability of specimens determines the capability of the control system to detect changes with time. In different specimens, the same analyte may exhibit different degrees of stability. Different analytes in the same specimen may exhibit different degrees of stability. Thus, the useful life of the material may depend on its application. Microbiological contamination can affect stability.

4.3.1. Stability should be established, by appropriate testing, both for the material as normally received and stored, and after any reconstitution which is required before use. Stability should be expressed as the shelf-life for each component under stated storage conditions.

4.3.2. The turbidity of the lyophilized material after reconstitution is often an analytical problem. The degree of turbidity can vary between different materials, and different batches of the same material. It can be affected by the time and conditions of storage.

4.4 Assignment of values

4.4.1. Primary calibration standards, i. e. those prepared from pure solute and pure solvent with concentration values established by preparation, have significant operational advantages over those materials whose values must be established by analysis (see Section 4.4.2), and they should be used wherever possible. The uncertainty level is only that associated with the operations of weighing and volume measurement and the limitations of purity evaluation. In many cases. this uncertainty is less than ± 0,2% of the mean value.

4.4.2. With secondary calibration standards, the uncertainty level is usually larger than that of primary standards, as it depends on the precision and accuracy of the method used to establish the value. Some complex biological secondary standards are prepared by adding weighed amounts of the analyte to a matrix from which most of the analyte has been removed. In these, the main uncertainty will arise from three sources: the error in weighing in the analyte, that of measuring the volume of fluid matrix base, and that arising from analysis of the matrix.

4.5 Uncertainty of designated values

The uncertainty of values determined by analysis should be calculated by appropriate statistical protocols. At least two recommendations have been published (12, 13), both of which involve multiple analyses by several independent laboratories, with replication, to reduce the confidence interval of the final value. The problem is neither simple nor trivial. Variations between and within laboratories involve both random variability and systematic bias. Systematic bias may occur between methods, between laboratories applying the same method, and in the same laboratory at different times. The definition of an effective protocol to assign values to products under these circumstances is a difficult statistical and operational challenge. Attempts to prescribe unique numerical values for acceptable variation or allowable error of results have no solid foundation (3); thus the desirable confidence intervals of calibration standards depend on the purpose for which the analytical method is to be used.

5. Preparation and handling

Difficulties arise in the preparation to desired specifications of specimens which resemble patients' specimens.

5.1

The starting materials used in the preparation of standards and control materials range from pure substances, essentially free of undefined components (e.g. 99,9% sodium chloride and pure water), through less pure sources (such as glucose with an undefined water content, or $MgCO_3$/$CaCO_3$ mixture), to incompletely defined analyte sources (such as human and animal serums and tissue extracts).

5.2

Selection of base material is largely a matter of informed judgement, based on a knowledge of matrix effects in the usual analytical processes, interference effects, desired analyte concentrations, stability factors, etc.

5.2.1. Human serum or plasma base material may be obtained from selected donors of whole blood or by plasmapheresis, but for ethical reasons this should be used in special cases only. Alternatively, surplus laboratory specimens can be used if they are collected without excessive microbiological contamination. Methods of preparing such specimens in the laboratory are described, for example, by *Bowers* et al. (1975).

5.2.2. Serum (or other biological fluid) from animals may also be used as base material. In some instances, the base material may be modified *in vivo* by appropriate treatment of the donor animal; e. g. for cholesterol and triglycerides (11), drugs, heavy metals, etc.

5.2.3. After collection of suitable base material, further processing may be necessary or desirable to ensure that the final product does in fact conform to analytical specifications. The following points may be of assistance:

1) Proper safety precautions should be observed in handling all serum and urine preparations, since these materials may contain hepatitis virus or other infective agents.

2) Base material obtained from surplus laboratory serum specimens should be kept at -20 °C or less during the period of collection and until required. Other supplies of base material, animal and human, should be stored under similar conditions.

3) Preliminary analysis of base materials is necessary to decide on the degree of modification required.

4) Individual specimens of base material can be selected into groups and pooled according to their analytical results, e. g. high and low cholesterol values. Graded concentrations of analytes may be obtained by careful combination of two or more such preparations. Some pools of material or individual patient sera must be considered for rejection when their properties are unusual or interfere in some analytical process (excessive lipaemia or haemolysis). The combination of base solutions with or without additives in relatively large volumes to achieve complete homogeneity is an art, and considerable time and care must be exercised to ensure complete mixing.

5) Base materials may be modified *in vitro* by a variety of other processes, some of which may be specific for a given analyte, whilst others are non-specific. For example, treatments such as selective membrane filtration will modify analyte concentrations and remove contaminating organisms, and treatment with enzymes will destroy specific analytes. Substances may be added to the base material either to achieve desired analyte concentrations or to increase stability. Broad spectrum antibiotics may be substituted for, or used in addition to, sterilization by filtration, provided that they do not interfere with the analytical method.

6) Although liquid base materials are most conveniently combined with, or modified by, other fluids, it is possible to add some materials quantitatively as solids. It should be remembered that, with soluble analytes, this will produce temporarily some pockets of high concentration which may destroy or modify other components. These difficulties will be minimized by rapid stirring of the solution until it is again homogeneous. The addition of detergents helps to disperse materials such as lipids in the biological matrix. The effect of these additives on the analytical process must be determined.

7) Clotting factor concentrations may be unpredictably affected by many treatment processes and create special problems. Also, incompletely coagulated serum, or mixtures of serum and plasma, will develop fibrin threads or clots, sometimes hours after processing is finished.

5.3

Whilst a preparation is stored in one or a few containers, it is generally not difficult to maintain sterility and freedom from ambient contamination. When the material is divided into multiple, small volume containers for despatch and use, the attainment of homogeneity and the maintenance of both sterility and freedom from contamination become much more difficult. Special attention must be paid to the compatibility of the container and its stopper with the liquid, dried, or reconstituted material and its intended use. Tests for sterility in the final container are relatively easy to carry out by allowing a number of the specimens to remain at room temperature for a few days to a few weeks. A developing turbidity is a good indication of microbiological contamination, as is a decrease in glucose content. Ambient chemical contamination can only be assessed by analysis as part of quality control during manufacture. This should include tests on empty containers.

5.4

Lyophilization greatly increases the stability of many constituents, particularly if the specimens produced are stored at low temperatures (4 °C or less).

5.4.1. The same volume of liquid specimen must be delivered into each individual container. Variation in this volume will result in differences in the concentrations of analytes in the reconstituted specimens.

5.4.2. To ensure minimal variability and predictable accuracy in lyophilized specimens, the exact conditions of lyophilization must be carefully controlled to achieve a solid residue with low residual moisture content and with a physical consistency which allows easy reconstitution.

5.4.3. Normally a stated volume of water is used to reconstitute the lyophilized material, but some manufacturers provide special solutions to reconstitute their products. In every case, the stated values of the reconstituted specimens are only valid if the manufacturers' recommended procedure is followed.

5.4.4. Analyte concentrations in the reconstituted specimens can be changed from the stated values by either or both of the following:

1) Reconstitute with more or less than the volume of fluid specified. It should be recognized that this may result in errors, if a significant matrix effect is present in the analytical process. The analyte concentration can be calculated in such cases only if the volume of dried material in the vial ("partial specific volume") is known,

or if the final reconstituted volume of the specimen is measured.

2) Reconstitute using an analyte solution instead of pure water. Again, it is important to be able to measure the final reconstituted volume or to calculate it.

5.4.5. With some analytes, a period of time (up to several hours) may be needed after reconstitution before a satisfactory analysis can be undertaken (Ca, Mg, certain enzymes). This is also true of frozen specimens whose homogeneity and stability must be established before use.

6. Bibliography

1. Büttner, J., Borth, R., Boutwell, J. H., Broughton, P. M. G. & Bowyer, R. C. (1977), Provisional Recommendation on Quality Control in Clinical Chemistry. Part 3. Calibration and Control Materials. Clin. Chim. Acta *75*, F11–F20; J. Clin. Chem. Clin. Biochem. *15*, 233–238 (1977), Clin. Chem. *23*, 1784–1789 (1977).
2. Büttner, J., Borth, R., Boutwell, J. H., Broughton, P. M. G., & Bowyer, R. C. (1978), Provisional Recommendation on Quality Control in Clinical Chemistry. Part 5. External Quality Control. Clin. Chim. Acta *83*, 189F–202F, J. Clin. Chem. Clin. Biochem. *16*, 259–266 (1978), Clin. Chem. *24*, 1213–1220 (1978).
3. Büttner, J., Borth, R., Boutwell, J. H., & Broughton, P. M. G. (1979a), Approved Recommendation on Quality Control in Clinical Chemistry. Part 1. General Principles and Terminology. Clin. Chim. Acta *98*, 129F–143F. J. Clin. Chem. Clin. Biochem. *18*, 69–77 (1980).
4. Büttner, J., Borth, R., Boutwell, J. H., Broughton, P. M. G. & Bowyer, R. C. (1979b), Approved Recommendation on Quality Control in Clinical Chemistry. Part 2. Assessment of Analytical Methods for Routine Use. Clin. Chim. Acta *98*, 145F–162F. J. Clin. Chem. Clin. Biochem. *18*, 78–88 (1980).
5. Büttner, J., Borth, R., Boutwell, J. H., Broughton, P. M. G. & Bowyer, R. C. (1980a), Provisional Recommendation on Quality Control in Clinical Chemistry. Part 6. Quality Requirements from the Point of View of Health Care. J. Clin. Chem. Clin. Biochem. *18*, 861–866.
6. Büttner, J., Borth, R., Broughton, P. M. G. & Bowyer, R. C. (1980b), IFCC Document with a view to an IFCC Recommendation on Quality Control in Clinical Chemistry. Part 4. Internal Quality Control. J. Clin. Chem. Clin. Biochem. *18*, 535–541.
7. Bowers, G. N., Burnett, R. W., & McComb, R. B. (1975), Selected Method: Preparation and Use of Human Serum. Control Materials for Monitoring Precision in Clinical Chemistry. Clin. Chem. *21*, 1830–1836.
8. Cali, J. P. (1975), International Critera for Reference Materials. In: Quality Control in Clinical Chemistry. Transactions of the 6th International Symposium, Geneva, April 23–25, 1975 (Anido, G., van Kampen, E. J., Rosalki, S. B., & Rubin, M., eds.) W. de Gruyter, Berlin–New York.
9. Cooper, G. R. (1969), Quality Performance in Clinical Pathology. Progr. Clin. Pathol. *3*, 1–71.
10. Eldjarn, L. (1968), Control Solutions in Clinical Chemistry, Z. Anal. Chem. *243*, 766–770.
11. Hollender, A. (1970), A Simple Method for Concentrating Serum by Freezing and Thawing. Scand. J. Clin. Lab. Invest. *25*, 63–64.
12. Moinuddin, M., Witter, R. F., & Quist, K. D. (1972), Bovine Blood Serum as a Substitute for Human Serum for Quality Control of the Determination of Cholesterol and Triglyceride. Clin. Chim. Acta *37*, 123–130.
13. NCCLS (1974) Standards for Calibration, Reference and Control Materials in Clinical Chemistry, National Committee for Clinical Laboratory Standards, Los Angeles.
14. Stamm, D. (1974), Calibration and Quality Control Materials. Z. Klin. Chem. Klin. Biochem. *12*, 137–145.
15. Stevens, J. F. (1973), Control Materials for Clinical Biochemistry. Tech. Bull. No. 29 Ann. Clin. Biochem. *10*, 133–145.
16. Whitehead, T. P., Browning, D. H. & Gregory, A. (1973) A Comparative Survey of the Results of Analysis of Blood Serum in Clinical Chemistry Laboratories in the United Kingdom. J. Clin. Pathol. *26*, 435–445.
17. Young, D. S. & Mears, T. W. (1968), Measurement and Standard Reference Materials in Clinical Chemistry. Clin. Chem. *14*, 929–943.

Part 4. Internal Quality Control

This paper is based on a Provisional Recommendation (*Büttner* et al., 1980c (9)) which has been revised to take account of the comments received, and has been accepted by the Council of the International Federation of Clinical Chemistry by mail ballot on 1983-01-31.

Contents

1. Introduction

1.1

For practical reasons such as operation, experimental design, and training of staff, it is convenient to distinguish between internal and external quality control (QC) (*Büttner* et al. 1979a (5)). The fundamental premise of both procedures is that errors detected by means of control specimens exactly mirror errors occurring with patients' specimens. Results are evaluated by appropriate statistical methods. The term "internal quality control" refers to procedures which are designed to be carried out within a laboratory for the purpose of monitoring the performance of that laboratory alone. In contrast, external quality control is a joint venture of two or more laboratories (*Büttner* et al. 1983 (10)). Although the term "statistical quality control" is sometimes used as a synonym for "internal quality control", it should be obvious that all comprehensive QC procedures include statistics as a tool.

1.2

In view of the many factors to be controlled, and subject to the size and organization of the laboratory, a variety of techniques are used in internal QC and have to be combined into an integrated effective system. The factors to be controlled include not only the analytical methods themselves, but also laboratory instruments; reagents and chemicals, including water; cleaniness of containers and other equipment; aspects of laboratory management such as work schedules, work flow, data transcription and communication; skill and training of staff, storage and handling of specimens (including health safety risks), and collection and transport of specimens.

2. Objectives

2.1

The general objective is to ensure the reliable and efficient functioning of the laboratory so that, in particular, valid results are returned early enough to influence medical decisions.

2.2

The following specific aims must receive attention:

(1) The primary aim is to maintain for all methods, including those used for emergencies, an analytical quality which should be specified taking both medical and analytical aspects into consideration. QC can, however, also serve as an aid for long-term improvement of performance.

(2) To alert staff so that the reporting of erroneous results is prevented and remedial action taken immediately.

(3) To ensure proper collection, transport, storage, and sampling of specimens where these are the laboratory's responsibility.

(4) To detect undue variation and identify its sources. This information can lead to improved performance through modifications in equipment, methods, organization, and training.

(5) To contribute to better patient care through improved diagnostic specifity and sensitivity (*Wilson & Jungner* 1968 (23)).

3. Limitations

3.1

The fundamental premise, i. e. the assumption that control specimens behave exactly like patients' specimens in the analytical system, may not be met. Potential causes include differences in matrix (e.g., control specimens of non-human origin), processing (mixing, filtration, dialysis, lyophilization), packaging (filling errors), and reconstitution (pipetting errors, handling).

3.2

Errors which are not detected by means of control specimens may arise in individual patients' specimens due, for instance, to the presence of interfering substances, whether endogenous (e.g., uremia) or exogenous (e.g., drugs or their metabolites).

3.3

Because of difficulties in assigning true values to control specimens, these are of limited value in measuring inaccuracy.

3.4

Open or semi-blind processing of control specimens (see Section *5.3*) may produce biased (over-optimistic) measures of performance.

3.5

Internal QC may be of little or no help in identifying the nature and source of some errors – for instance, those due to collection, transport, or storage of patients' specimens.

3.6

If the same staff members are responsible both for the process to be controlled and for the control procedure, objectivity may be impaired, and remedial action delayed or prevented.

3.7

The correct use of internal QC requires a basic knowledge of statistics from all operators involved. Moreover, the statistical criteria most widely used involve probabilities of chance occurrence derived from Gaussian distributions, but other distributions may occur in practice.

3.8

It is difficult to verify that an internal QC procedure is effective. To make a QC system more comprehensive requires considerable expenditure, the benefit of which may not be immediately apparent.

4. Materials

4.1

Specimens suitable for use in internal QC (*Büttner* et al. 1980a (7)) may be either in natural form (such as material from patients), or specially prepared. The general rule is that control specimens should be as similar as possible to the material being examined with respect to composition and physical characteristics such as viscosity and density – that is, they should have a similar matrix.

4.2

In practice, the following materials are used:

(1) Surplus specimens from patients, either pooled or used individually. The material may be unstable and infectious, and may contain interfering factors. It has, however, the advantage that matrix effects can be minimized since the control material can be expected to contain the same interfering factors as the specimen. The material is cheap and readily available and can be kept stable by sterile filtration and storage below −70 °C.

(2) Specially prepared control specimens are more stable and can be made up to a more completely defined composition (*Büttner* et al. 1980a (7)).

4.3

Special materials and devices (dye solutions, test filters, absorbancy standards, etc.) are required for the performance control of measuring instruments. They are not dealt with in this document.

4.4

Aqueous solutions of pure analyte should not be used since in this case the matrix differs greatly from that of the patients' specimens.

5. Internal QC procedures and data acquisition

5.1

The basic principle and prerequisite of all internal QC procedures is that control specimens should be treated in exactly the same way as patients' specimens.

5.2

Internal QC procedures are intended to reveal changes in performance from a carefully determined initial state. The initial assessment of analytical methods has been described (*Büttner* et al. 1979b (6)). In QC it is assumed that methods are subject to inevitable variation that can be monitored by means of statistical techniques. After any change in the method or conditions of analysis, it is mandatory to determine the initial state of performance again.

5.3

Various aspects of internal QC may call for different designs, for instance regarding the use of "open" or "blind" controls – a distinction which is of basic importance. Controls are called "open" if the analyst knows their identity and their assigned values. In a "blind" design, it is made impossible for the analyst to identify the control specimens. Designs are often "semi-blind", i.e., the analyst is aware of control specimens but does not know the expected values. Only blind designs obviate the analyst's conscious or unconscious bias, and ensure that the procedure is monitoring the same sources of error as those affecting patients' specimens. However, blind designs are often not feasible because control specimens and the associated request are difficult to disguise. Moreover, their use is bound to delay remedial action in the event of a control system alert. For these reasons, a dual system including both open and blind (or, at least, semi-blind) controls (*Tonks* 1968 (17)) is best suited to serve different needs: the open controls function as "bench controls" and without delay supply the analyst with information for immediate remedial action, while the blind controls provide unbiased information.

5.4

Accuracy and precision are the reliability criteria usually monitored (*Büttner* et al. 1979a (5)), but it can be helpful to include dependability and other criteria.

5.5

In the simplest design, analyses are done in duplicate on a series of different specimens, with the duplicate samples placed either in the same or different runs (see definition *Büttner* et al. 1979a (5)). The range of results within each pair provides information on imprecision alone. The design may be used with patients' specimens by simply repeating some analyses on, say, the next day, provided that the analyte is known to be stable.

5.6

More information is obtained from several replicate analyses of the same homogeneous control specimens. This design provides information not only about imprecision, but also about changes in inaccuracy. If control specimens with known values are used, inaccuracy is actually measured at the same time.

5.7

An evaluation of results of a number of patients' specimens (see Section *6.6*) may indicate changes in accuracy including those due to sampling, transportation, and preparatory treatment of specimens. No special material or effort is needed apart from data collection and calculation.

5.8

In many analytical methods, imprecision and inaccuracy are not constant over the analytical range and should be controlled for various concentrations or activities chosen to include points where clinical decisions are made (e.g. serum bilirubin at 340 µmol/l in paediatrics). In a comprehensive design, concentrations below and above the reference interval ("normal range") should be included. Costs for such designs may be excessive.

5.9

In cases where control materials are difficult to prepare and unstable, and where the biological variation in an individual is known to be small (e.g., blood gases and pH), it can be appropriate to use single specimens taken from subjects known to be in a stable state for monitoring changes in precision and accuracy.

5.10

There is no general rule about the type and number of control samples which should be analysed, as this will depend on the information required: imprecision within or between runs, imprecision and/or inaccuracy at several concentrations (see Section *5.8*), and on whether controls are to be open or blind or both. The minimum number of controls for all types of test (including stat or emergency tests) is one for every batch of patients' specimens, but the use of more controls will markedly reduce the risk of an "error of the second kind", i.e., of failure to alert when a fault is present (*Büttner* 1967 (3)). With mechanized equipment, special designs may be needed, such as including several samples of a control specimen within a run to measure drift, or specimens of markedly different concentrations to measure carry-over or sample interaction (*Hjelm* 1968 (13); *Broughton* et al., 1974 (2)). It must be emphasized that results obtained for such purposes of calculation or correction must not be used for control purposes.

5.11

It is useful to check whether the time between requesting a test and receiving the result (turn-around time) is within acceptable limits. A suitable design – which at the same time tests the organization of work flow – is to introduce control specimens as fictitious patients' specimens, for instance in the ward, and to record the times when a specimen is despatched and the laboratory report received.

5.12

Suitable organization is needed to make internal QC effective. A QC supervisor should be designated and given the responsibility for the selection of control material, experimental design, and evaluation of QC data. The control procedure should provide information on run quality so quickly that patients' results can be reviewed without delay before they leave the laboratory. There should be clear instructions and assignments of responsibility for remedial action when an alert occurs. Analysts should be kept informed of all QC results and conclusions, in order to maintain interest and promote long-term improvement.

6. Evaluation

6.1

The primary objective of internal QC is attained if the system gives an alert when an error occurs in an analytical run. Measures can then be taken immediately to remove the source of error.

6.2

In most procedures for internal QC, errors affecting a batch of patients' results are detected by examining the results of control specimens analysed in the same batch. These results are a "sample" in the statistical sense, and the analyst has to decide on statistical grounds (*Büttner* et al. 1979a (5)) whether the current sample is likely to belong to the same population of control results as those obtained when the initial state of performance was being defined (see Section *5.2*). This statistical test can be carried out by calculation, but it is better to involve the operator personally in plotting results graphically on a control chart (for the use of patients' results in internal QC, see Section *6.6*).

6.3 Control charts

6.3.1. In statistical terms, the hypothesis being tested is that the mean result found for the control specimen is not different from the expected value. The test statistic used is the deviation from the expected value divided by its standard error.

$$\xi = | \bar{x} - \mu_0 | / (\sigma_0 / \sqrt{N})$$

where $\bar{x}$ is the mean of N replicate assay results for the specimen, μ_0 the given target value, and σ_0 the standard deviation of replicate results, chosen for the process. If ξ exceeds a value λ, the hypothesis is rejected, i.e. the sample does not meet the quality requirements which have been set. By the choice of λ, the probability of false alerts (statistical errors of the first kind) can be kept as small as desired – for example, in the case of the Gaussian distribution, at 1/20 with $\lambda = 2$, or at 1/100 with $\lambda = 3$.

6.3.2. The control chart is used to test whether, over a prolonged period of time, results of control specimens vary more widely than expected by chance; in other words, whether no other than random errors are present. It is therefore essential that the standard deviation used to calculate the control limits is derived from replicate analyses over an appropriately long time, e.g. day-to-day variation over 3–4 weeks.

6.3.3. The principles underlying control charts have been described (*Büttner* et al. 1979a (5)). Time or a time-dependent quantity (date, run or sample number) is plotted as abscissa, the quantity measured (concentration, activity per unit volume, counts per minute) as ordinate. This type is called a "single value control chart". Statistical reliability can be enhanced by plotting a statistic derived from a set of values such as the mean result of several control samples ("mean value control chart"), standard deviation, coefficient of variation, or range.

6.3.4. If the error distribution characterized by μ_0 and σ_0 of the initial state is unchanged, 95% ($\lambda = 2$) or 99% ($\lambda = 3$) of the plotted points can be expected to lie between the decision limits ("warning" and "action" control limits, respectively) which, if the plotted points are means, are defined as

$$L = \mu_0 \pm \lambda \cdot \sigma_0 / \sqrt{N}$$

The previous estimate of variation used (σ_0) should preferably be based on at least 20 to 30 values measured in different runs (*Büttner* et al. 1979a (5)).

6.3.5. A plotted point outside the control limits indicates (because the probability of this occurring through chance is smaller then 0.05 or 0.01) that precision or accuracy or both have probably changed from those of the initial state. This constitutes an alert, and the analytical method is said to be "out of control".

6.3.6. The user should be aware of the limitations of control chart techniques. One is the assumption, underlying most statistical techniques in current use, that the distribution of the variate tested is Gaussian ("normal"). There is some evidence that other distributions (Poisson, log-normal) in certain cases better describe the data. Another important limitation is the interdependence of errors of the first and second kind. The probability of errors of the first kind (false alert) is fixed, usually at 0.05 and 0.01 as explained, by setting the control limits. The probability of errors of the second kind (no alert although analytical method out of control) depends in a complex manner on the distribution assumed for out-of-control values, the control limits set, and the number of analyses in the run. The relations can be explored by calculating the "operating characteristic curve" of the control chart (*Büttner* 1967 (3)).

6.3.7. Control charts are not as sensitive to changes within decision limits, and therefore not as reliable in revealing trends, as are methods such as the cusum control chart (*van Dobben de Bruyn* 1968 (18)). Alternatively, according to the theory of runs (*Wolfowitz* 1943 (24)), a method is likely to be out of control if more than 7 successive points lie one side of the central line or occur in increasing or decreasing order.

6.4

Imprecision is defined as the standard deviation or coefficient of variation of the results in a set of replicate measurements (*Büttner* et al. 1979a (5)). From appropriate sets of data, imprecision can be calculated between or within runs or days.

6.5

The control of inaccuracy requires a statistical comparison of the results obtained for a control specimen with the target (true or assigned) value, using the previously determined imprecision of the method (see *Büttner* et al. 1979a (5)). The test can be made by calculation, or using a single or mean value control chart (see Section *6.3*).

6.6

Provided that a sufficient number of results is available, and assuming that the population of patients is stable, changes in the daily average of all patient results indicate changes in inaccuracy (*Dixon & Northam* 1970 (11); *Hoffmann & Waid* (14); International Symposium 1968 (15)). Some versions of this method use a truncated distribution – for instance, only results lying within the reference interval. The variation in the daily mean includes errors due to collection, transportation, and preparatory treatment of specimens. However, uncertainty about the stability of the patient population and the time and effort of calculation have relegated these methods to auxiliary use.

6.7

It is essential that QC data are presented effectively and displayed in a way which facilitates rapid interpretation, decision-making, and education. This is best achieved by displaying control charts close to the area where the analyses are performed. If charts show a derived statistic such as range or standard deviation, the original data (say, absorbance readings) should be readily available for inspection. Charts should include a record of events likely to influence performance, such as new calibration standards or repair of instruments.

6.7.1 Control analyses may amount to as much as 20% of the work-load of the laboratory, and the large volume of data produced may need to be reduced to facilitate interpretation. One way of achieving this (which is made easier by a computer) is to calculate simple indices of performance, such as a precision index and an accuracy index (*Whitehead* 1977 (21)), which may be sufficient for error detection. More detailed QC information, including the original data, can be examined as necessary to investigate the type of error.

6.8

Internal QC data are of relevance in the correct medical interpretation of laboratory results and should be presented to clinicians in an appropriate manner (e.g., confidence limits).

7. **Remedial action**

7.1

Because QC data can be made available as soon as an analysis is completed, a well-run internal QC system makes possible immediate intervention in the release of patients' results. This advantage cannot be overrated. All batches of all analyses can and should be monitored by internal QC – a coverage which, for practical reasons, is not achieved by external QC programmes.

7.2

In the event of a control system alert, it is advisable to proceed in order through the following steps:

(1) Decision: immediate decision whether action is necessary.

(2) Investigation: checks to locate the error.

(3) Repair: action to eliminate the error.

Details of this scheme will depend on the method and equipment involved as well as other circumstances. Therefore, only general guidelines can be given.

7.3

In deciding whether or not to take action, one should recognize that statistical tests admit a stated probability of false alerts (see Section *6.3*). Also, one should consider whether a single alert has occurred after a trend in control values, or whether it involves only a particular control specimen while results from other specimens continue to be acceptable. The practice of merely repeating the analysis of those control specimens outside the warning limits and hoping for the new values to be within those limits completely alters the statistical basis of the control chart and should not be followed.

7.4

If the decision is taken that the method is out of control, the first action is to withhold patients' results. Then the search for the source of the error begins. The analyst should start by checking for the simplest and most frequent faults, and then continue as necessary in a logical order depending on the method and equipment involved. The following sequence is suggested.

7.4.1. It is good practice to start by excluding gross errors such as mix-up of specimens, reagents or pipets, misuse of measuring instruments (wrong filter or lamp in the photometer), or failure to follow instructions for a step in the method.

7.4.2. The result obtained on a control specimen may be in error for several reasons including deterioration due to age, incorrect storage, or contamination; wrong identification; mistakes in preparation or reconstitution.

7.4.3. If no fault has been found up to this point, further action will depend on whether the alert is due to a change in accuracy or in precision. If inaccuracy has increased, attention should be focussed on the possibility of systematic sources of error such as incorrect reaction temperatures, calibration errors, and faulty devices. If precision has deteriorated, steps in the analytical procedure should be checked – for instance, deproteinization, dialysis, composition of reagents and reaction mixtures, measuring systems (photometer, flame photometer).

7.4.4. If the error source has still not been found, all operations in the method should be observed carefully, paying attention to details such as exact timing, correct reaction temperature, precise measuring, correct mixing, occurrence of turbidity.

7.5

Once a source of error has been identified, it can usually be eliminated fairly easily. A suitable record of the event should be kept.

7.6

Usually all specimens in a faulty batch will have to be analysed again after removal of the error source, unless the latter was due solely to the control specimen used.

8. Long-term aspects

8.1

Internal QC is more than just a warning system for monitoring analytical methods: it can and should serve as an aid to long-term improvement in performance. If used effectively, the QC data of a laboratory will provide much information that is valuable for the continuing education and training of staff, the improvement of analytical methods, the effective use of equipment, the assessment of different QC procedures, and for improvements in laboratory organization. Moreover, the QC data can provide laboratory staff with up-to-date assessments of the reliability of their work.

8.2

To achieve these aims, it is recommended that QC reports be prepared regularly on all analytical methods used, including emergency and stat tests. Ideally, the reports should include the following items:

(1) Average imprecision and average inaccuracy of all methods over the stated period.

(2) Other information about performance, such as drug interferences encountered, turn-around time for various tests.

(3) Number and type of out-of-control events detected, with date and other relevant details.

(4) Assessment of the efficiency of different QC procedures (including external QC programmes) in detecting errors.

(5) Details of the control materials used, and any problems encountered with them.

(6) Causes of error which were identified and removed by remedial action. This may be particularly important in calling attention to needs for maintenance or replacement of equipment, or for additional training of staff.

(7) Comments from clinicians on errors which may have escaped detection, and on other needs for improved performance.

8.3

Regular staff meetings should be held to discuss these aspects critically, and to consider suggestions for improvement. It should be stressed and repeated that improvement of performance is a slow process which can only be accomplished by continuing hard work and the cooperation of all involved.

References

1. Broughton, P. M. G. & Raine, D. N. (1969), Quality control in clinical chemistry. Ann. Clin. Biochem. *6*, 89–148.
2. Broughton, P. M. G., Gowenlock, A. H., McCormack, J. J. & Neill, D. W. (1974), A revised scheme for the evaluation of automatic instruments for use in clinical chemistry. Ann. Clin. Biochem. *11*, 207–218.
3. Büttner, J. (1967), Statistische Qualitätskontrolle in der klinischen Chemie. Z. Klin. Chem. Klin. Biochem. *5*, 41–48.
4. Büttner, J. Hansert, H. & Stamm, D. (1974), Statistical analysis, control and assessment of experimental results. In: Methods of Enzymatic Analysis, (Bergmeyer, H. U. ed.) Weinheim, Vol. 1, 318–395.

5. Büttner, J., Borth, R., Boutwell, J. H., & Broughton, P. M. G. (1979a), Approved Recommendation on Quality Control in Clinical Chemistry. Part 1. General Principles and Terminology. Clin. Chim. Acta *98*, F129–F143; J. Clin. Chem. Clin. Biochem. *18*, 69–77 (1980).
6. Büttner, J., Borth, R., Boutwell, J. H., Broughton, P. M. G. & Bowyer, R. C. (1979b), Approved Recommendation on Quality Control in Clinical Chemistry. Part 2. Assessment of Analytical Methods for Routine Use. Clin. Chim. Acta *98*, F145–F162; J. Clin. Chem. Clin. Biochem. *18*, 78–88 (1980).
7. Büttner, J., Borth, R., Boutwell, J. H., Broughton, P. M. G. & Bowyer, R. C. (1980a), Approved Recommendation on Quality Control in Clinical Chemistry. Part 3. Calibration and Control Materials. J. Clin. Chem. Clin. Biochem. *18*, 855–860.
8. Büttner, J., Borth, R., Boutwell, J. H., Broughton, P. M. G. & Bowyer, R. C. (1980b), Approved Recommendation on Quality Control in Clinical Chemistry, Part 6. Quality Requirements from the Point of View of Health Care. J. Clin. Chem. Clin. Biochem. *18*, 861–866; Clin Chim. Acta *106*, 109F–120F.
9. Büttner, J., Borth, R., Broughton, P. M. G. & Bowyer, R. C. (1980c), IFCC Document with a view to an IFCC Recommendation on Quality Control in Clinical Chemistry. Part 4. Internal Quality Control. J. Clin. Chem. Clin. Biochem. *18*, 535–541.
10. Büttner, J., Borth, R., Boutwell, J. H., Broughton, P. M. G. & Bowyer, R. C. (1983), Approved Recommendation (1983) on Quality Control in Clinical Chemistry, Part 5. External Quality Control. J. Clin. Chem. Clin. Biochem. *21*, 885–892.
11. Dixon, K. & Northam, B. E. (1970), Quality control using the daily mean. Clin. Chim. Acta *30*, 433–461.
12. Fraser, C. G. & Peake, M. J. (1980), Problems associated with clinical chemistry quality control materials, CRC Critical Reviews in Clinical Laboratory Sciences, 59–86.
13. Hjelm, M. (1968), Quality control of automated analytical systems in clinical chemistry. Z. Anal. Chem. *243*, 781–790.
14. Hoffmann, R. G. & Waid, M. E. (1965), The "average of normals" methods of quality control. Amer. J. Clin. Pathol. *43*, 134–141.
15. Internal Symposium on Statistical Quality Control in the Analytical Laboratory (1968), Z. Anal. Chem. *243*, 751–825.
16. Levey, S. & Jennings, E. R. (1950), The use of control charts in the clinical laboratory. Amer. J. Clin. Pathol. *20*, 1059–1066.
17. Tonks, D. B. (1968), A dual programme of quality control for clinical chemistry laboratories, with a discussion of allowable limits of error. Z. Anal. Chem. *243*, 760–765.
18. Van Dobben de Bruyn, C. S. (1968), Cumulative Sum Tests: Theory and Practice. Griffin's Statistical Monographs and Courses. Vol. 24. Griffin. London.
19. Westgard, J. O., Falk, H. & Groth, T. (1981). A multi-rule Shewhart chart for quality control in clinical chemistry. Clin. Chem. *27*, 493–501.
20. Whitby, L. G., Mitchell, F. L. & Moss, D. W. (1967), Quality control in routine clinical chemistry. Adv. Clin. Chem. *10*, 65–156.
21. Whitehead, T. P. (1977), Quality Control in Clinical Chemistry, Wiley, New York.
22. Whitehead, T. P. (1977), Advances in quality control. Adv. Clin. Chem. *19*, 175–205.
23. Wilson, J. M. G. & Jungner, G. (1968), Principles and practice of screening for disease. WHO Public Health Papers No. 34, WHO, Geneva.
24. Wolfowitz, J. (1943), On the theory of runs with some applications to quality control. Ann. Math. Stat. *14*, 280–288.

Part 5. External Quality Control

This paper is based on a Provisional Recommendation (*Büttner* et al., 1978 (1)) which has been revised to take account of the comments received, and has been accepted by the Council of the International Federation of Clinical Chemistry by mail ballot on 1983-01-31.

Contents

1. Introduction

1.1

External quality control is a procedure for utilizing, for quality control (QC) purposes, the results of several laboratories which analyze the same specimen(s) (*Büttner* et al. 1979a (2)). Many different types of scheme are in use throughout the world, varying in scope from simple arrangements, involving the exchange of specimens between two laboratories for analysis of a single component, to extensive national and international programmes, involving hundreds of laboratories measuring many components, and relying on computer analysis of the results. These larger schemes have been organized by commercial interests, professional bodies, government agencies, and as cooperative ventures between two or more of these groups.

1.2

In some countries, laboratories are required by law to participate in external QC schemes, and their results are inspected by a controlling or accrediting agency. In others, participation is voluntary, and the results of individual laboratories may only be known to the organizers or, by agreement, to other participants or interested persons. Opinions differ on whether remedial action is more likely to be taken if participation is voluntary or required by law, and anonymous or not.

1.3

External QC has been referred to by a number of other terms, e.g. proficiency testing of laboratories, interlaboratory QC, external surveillance, surveys, roundrobins, *Ringversuch*.

2. Objectives

The aims of external QC are:

2.1

To provide a measure of the "state of the art" for a test. Thus, the overall standard deviation (S.D.) of all results is a measure of the ability of different laboratories to obtain the same result on the same specimen. If the true value of the specimen is known, the results also provide an overall measure of inaccuracy.

2.2

To provide a measure of the quality of performance of individual laboratories, which can be compared with that of other participants, with that of selected laboratories, or with defined standards of performance. By regular testing, changes in the quality of performance can be studied.

2.3

To supplement internal QC procedures within a laboratory.

2.4

To obtain consensus values for specimens which can be used in the assessment of analytical methods, when true values are not known (see *Büttner* et al., 1979b (3) (Section *6.9.2), Gilbert* 1978 (10)).

2.5

To investigate features within a laboratory which may contribute to its performance and cannot be studied by internal QC procedures alone. These features include the analytical method used, the size and staffing of the laboratory, the work load, test frequency, etc. It is also possible to measure individual laboratory and group performance with respect to calibration and blank errors, materials stability and the drift in values with time, and to investigate reagent problems and defects in internal QC. Each of these requires careful planning and appropriate experimental design.

2.6

To act as an educational stimulus to improvement of performance. In some cases it is possible, directly or indirectly, to suggest specific remedial action to laboratories producing unsatisfactory results.

3. Advantages and limitations

3.1

Since specimen values are not known at the time of analysis, the results should provide an objective measure of performance. As with all QC methods, laboratories must not give control specimens special treatment, and methods of reducing this temptation should be considered when designing the programme. This is particularly important when participation is required by law, as there will then be additional pressure to obtain the best possible results for these specimens, thus not representing the laboratory's usual performance.

3.2

External QC is generally not a substitute for internal QC because testing is too infrequent for adequate within-series monitoring (but see Section *3.8*). Neither should it be used as the accuracy base of the laboratory whereby results from QC material are used to determine a correction factor to be applied to results obtained on patients' specimens. The practice of using the results of any control specimens to alter the values of calibration standards should not be tolerated.

3.3

Many programmes have as participants only those who voluntarily choose to take part. Such a selection of well motivated and concerned laboratories can be assumed to include the more competent performers, and their results will not be representative of all laboratories. Even when cost is not a factor, laboratories which are most in need of QC may be the most reluctant to participate.

3.4

For these reasons, in many schemes the identity of the participants is known only to the organizers. This is intended to encourage participation by less proficient laboratories, avoid adverse publicity, and protect professional reputations. On the other hand, disclosure of identity helps laboratories to discuss their results frankly, and enables the profession to see which laboratories are most in need of help. However, there are degrees to anonymity, and some safeguards are necessary to prevent unjustified public pillory of poor performers. Disclosure must be by consent.

3.5

It is important that the scientific basis of an external QC scheme, and the method for analyzing the results, are made clear to all participants, so that they have confidence in the results. Without such information, laboratories which participate in several different schemes may find the results conflicting and difficult to relate to their internal QC results.

3.6

If information is available about methods, standards, equipment etc., it becomes possible to allocate results to defined sub-groups (e.g. of analytical method). This is particularly important if methods are known to give different results. The number of results within each subgroup must be sufficiently large if reliable comparison is to be made between intragroup and inter-group performance and between "method means" (mean results obtained by different analytical methods). The reliability of method means is of particular importance when they are taken as target values (Section *6.2*).

3.7

Organizers must give clear instructions as to how participants should describe their methods. A general description (e.g. flame photometry or glucose oxidase method) will hide innumerable variations, and any conclusions drawn about the quality of the methods concerned may not be valid. However, the degree of complexity in the method description must be governed by the number of subgroups which can be handled, as discussed in Section *3.6*.

3.8

If a major aim of the scheme is to supplement internal QC (e.g. for a particular problem between two laboratories), it is essential that results are returned to participants rapidly enough to influence the reporting of patients' results. If they are used for survey purposes, they will lose their impact progressively if they are delayed, and may no longer reflect the current situation.

3.9

Care must be taken in assessing the reliability of a method in which the control specimens used do not behave in exactly the same manner as patients' specimens.

3.10

No scheme can be better than the quality of the control specimen used. Where this is a lyophilized material, results will reflect both analytical variation and errors in manufacture and/or reconstitution. The latter may contribute significantly to the overall variability of results.

3.11

Small local schemes are simple to organize and require no special equipment, and participants can meet to discuss the results. They have the added advantage that they can use specimens from individual patients ("split specimens"), and study components of limited stability. Concentrations unachievable by artificial fortification or depletion, and interfering compounds in their natural matrix, can also be included. However, asssessment of accuracy with such specimens of limited stability is usually difficult or impossible.

4. Materials

4.1

This general topic is discussed in Part 3 of this Recommendation (*Büttner* et al., 1980a, (4)). The following points are of particular relevance to external QC.

4.2

In order to generate confidence in the results, and especially to stimulate individual laboratories to remedy any revealed defects, it is vital that participants have confidence that specimens are stable, exhibit minimal between-vial variation, and are suitable for the method being used. It is, therefore, important to state the nature of the specimen and any limitations in applying it to particular analytical methods. If any doubt about the quality or the suitability of the specimen exists, participants will tend to blame this, rather than their analytical technique, for any poor results.

4.3

The volume of specimen required by each laboratory will depend on the methods used, and where a number of tests are to be performed on one specimen, the volume required may vary considerably. However, in the interest of economy, not more than 10 ml of serum is usually supplied except in surveys covering a large number of analytes. When the volume is insufficient, the individual participant has to decide which tests will have to be omitted, and he may exclude those which he performs less satisfactorily. To provide too large a volume of specimen will, on the other hand, encourage unwarranted replicate analyses.

4.4

Clear and detailed instructions should be given on the preparation of the material for analysis, particularly if it is lyophilized. Information on stability, recommended storage conditions, and minimum and maximum span of time before analysis of reconstituted materials should be included.

4.5

Materials which are appropriate to the survey design must be selected, and an adequate supply arranged. Laboratory testing may be essential at this time to ensure that the materials conform to expected standards of sterility and between-vial reproducibility, do not contain hepatitis associated antigen, and will withstand exposure to the conditions likely to be encountered during transport.

5. Survey design

5.1

It is essential to recognize that the information which can be obtained from external QC is directly dependent on the survey design used (*Stamm* 1971 (17)). Many different designs have been used but there can be no one "approved design" when the requirements and priorities of clinical laboratories, health administrations, and government agencies vary so much from place to place. The detailed features of a programme are obviously dictated by local requirements, but there are broad areas common to all.

5.2

A preparatory phase is needed for the selection and documentation of the factors which are to be studied, e.g. methodology, method of calibration, staff numbers, work-load, instrumentation. At this stage it is essential to ensure that specimens will arrive at all destinations within the allotted time. This will require correct addresses, and a careful appraisal of alternative transport systems. Where specimens have to be shipped across national borders, the necessary customs and quarantine clearances must be obtained prior to despatch.

5.3

The analysis phase may be complex or simple in design, depending on the aims of the programme. The following points should be considered.

5.3.1 If possible, survey specimens should be treated within the laboratory in exactly the same manner as patients' specimens. However, it is probable that specimens will be processed "semi-blind", i.e. the analyst is aware of the special nature of the specimen, but does not know the expected value. For the reasons outlined earlier (Section *3.1*), the laboratory may be tempted to give the specimen special treatment, such as replicate analysis, use of a special method or analyst, or even a special laboratory to analyze external QC specimens. In addition, two participating laboratories may cooperate informally and report a mutually agreed result. These temptations become greater when specimens are unnecessarily large and results need not be reported rapidly. While it is impossible to prevent special treatment, and difficult to prevent replicate testing, it is possible to reduce collusion by a simple randomization of distribution of various specimens within and between shipments.

5.3.2. If the design uses a number of replicate analytical results from each laboratory, an assessment can be made of within-laboratory precision and of differences in accuracy between laboratories. This cannot be completed until the results of a sufficient number of replicates have been returned, which may take several weeks. If the design includes a set of specimens with known interrelated values, it is possible to evaluate the type of bias present (e.g. proportional, constant or combined) (*Caragher* et al., 1977 (7)).

5.3.3. Assessment of accuracy depends on comparison of a mean value of a group of results with its appropriate target value, which can either be determined before the specimen is sent to participants, or can be based on participants' results. Methods of deriving target values are discussed below (Section *6.2*).

5.3.4. For the most effective between-laboratory evaluation, analyte concentrations should be used which span the clinically important range. Both imprecision and inaccuracy in a laboratory can vary with concentration, and it is important to establish that satisfactory performance is being achieved, not only within the reference interval, but also above and below it.

5.3.5. It is useful to include calibration standards in the programme to allow detection of calibration errors. Such standards may include pure solutions applicable to the methods employed by participants.

5.3.6. The frequency of testing will, in many cases, be determined on economic grounds, and the practical problems of sending out specimens, analyzing the results and returning these to participants promptly. Obviously, the greater the frequency, the greater the benefit.

5.4

After the specimen has been analyzed, the results must be returned to the organizers within a specified time. This should be as short as possible, so that the analysis of results can be sent to participants in time for them to take any necessary remedial action. The reporting operation can be a source of major errors in QC surveys as well as with patient results. These will be more prevalent where laboratories report results in units different from those demanded by the survey organizers. Well designed report forms will minimize the likelihood of mistakes, but are no substitute for care, legibility and attention to instructions.

6. Evaluation of results

6.1

The purpose of this stage is to obtain information about the performance of individual laboratories and groups of laboratories, and to present this as clearly and promptly as possible. Basically two calculations are required:

(a) A measure of the difference between an individual result and the target value for that specimen, and the relation of this to a performance standard.

(b) A measure of the location and dispersion parameters of grouped results.

6.2

Target values can be defined in several ways:

6.2.1. As the mean value obtained by a reference method ("reference method mean") or a method of known (small) bias; outliers (see *6.3*) may or may not be removed.

6.2.2. By using specimens with values established by the preparation process, e. g. by weighing in a known amount of salicylate to a serum known to contain no salicylate.

6.2.3. As the grand mean of all results after removal of outliers.

6.2.4. As the mean of a group of selected "reference" laboratories with known good performance ("reference laboratory mean").

6.2.5. As the mean of all results obtained by different laboratories using the same method, after exclusion of outliers („method mean").

6.2.6. As combinations of these, using all available information to arrive at a single target value (see "consensus value", *Büttner* et al., 1979b, (3), Section *6.9.2*)). In each case it is essential that the method of deriving the target value is clearly defined and validated, and is made known to participants.

6.3

"Outliers", i.e. discordant data not representative of the great majority of results, are due largely to arithmetical or clerical mistakes. They should therefore be identified, and subsequently omitted in the calculation of group mean and S. D., as they could otherwise exert gross influences on both of these (but see Section *6.3.2*).

6.3.1. A method frequently used to remove outliers is to truncate or eliminate all results lying outside the mean ±3 S.D. However, this simple treatment gives useful and reproducible results only where the distribution of valid results is Gaussian. Such a situation can never be strictly proven in practice, but it may be approximated in most cases, provided that only a few markedly deviant results are identified. An alternative approach is to use distribution-free percentiles, or to set arbitrary limits for truncation based on professional judgement.

6.3.2. The proportion of outliers may well vary between sub-groups, and provide important indications of the relative incidence of gross errors with, say, different methods, or in different types of laboratories. Statements of group performance should therefore specify whether outliers have been excluded, describe the criteria used, and state their proportion.

6.3.3. This information is also important when assessing the performance of individual laboratories. The inclusion of outliers will result in a measure of "total" laboratory performance. If they are excluded, the results are more likely to reflect normal analytical performance, and it could then be useful to state the proportion of outliers separately as an indication of gross errors. It is, however, up to the laboratory, not the statistician, to define what is meant by a gross error.

6.4

Performance standards for individual results may be based on:

6.4.1. Professional judgement, in the light of performance in other laboratories, regions, or nations.

6.4.2. The S.D. (or some other dispersion parameter) of all participating laboratories, or a relevant subgroup (say, all laboratories using the same method): for example, results within mean ± 2 S.D. may be rated as satisfactory, within the mean ± 2−3 S.D. as acceptable, and those outside mean ± 3 S.D. as unacceptable.

6.4.3. The S.D. (or some other dispersion parameter) of a group of "reference" laboratories; for example, acceptable limits may be set at mean ±3 S.D. of "reference" laboratory results.

6.4.4. Medical or diagnostic criteria: accuracy and precision limits may be defined in terms which relate to the biological relevance of the data, not to the current state of the analytical art.

6.5

External QC has always been regarded as a tool by which individual laboratories are stimulated to improve their performance. It is, therefore, of fundamental importance that the staff of all laboratories understand, from the summaries sent to them, how they have performed in relation to other laboratories. The information may be presented in several ways.

6.5.1. As a simple difference, positive or negative, from the target value in the units of the analysis, or as a percentage of the target value. Alternatively, the difference may be expressed in units of the S.D. of all results, or of the S.D. of a related group of results. Yet another variant is to report differences for all participants in ranking order, e.g. as a numerical frequency histogram, to enable each laboratory to compare its performance with that of others.

6.5.2. Many schemes make use of graphical displays such as frequency histograms and scatter diagrams, with defined performance limits, e.g. S.D. scales, clearly shown. The "Youden plot" (*Skendzel & Youden*, 1969 (16)) uses a graphical display of two

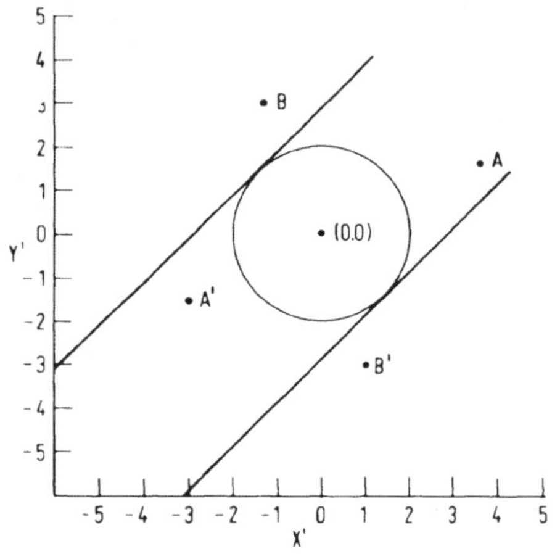

Fig. 1. Example of a Youden plot

unrelated specimens of different analyte concentration to illustrate, in a limited fashion, the presence of imprecision and inaccuracy, as shown in figure 1. Results on specimens X and Y from each laboratory are transformed by the relationships

$$x'_i = (x_i - \bar{x}) / S.D._x \text{ and } y'_i = (y_i - \bar{y}) / S.D._y$$

where $\bar{x}$ and $\bar{y}$ are the means of all x and all y, respectively, and $S.D._x$ and $S.D._y$ are the standard deviations of all x and all y, respectively (outliers excluded in each case). The transformed pairs of values are plotted on a diagram such as that shown. The S.D. scales on both x and y axes are identical, so that a circle may be drawn with centre 0,0 (transformed from $\bar{x}$, $\bar{y}$) and radius 2 S.D. which, in theory, should contain approximately 95% of results. Two parallel tangents, at 45° to both x and y axes, are drawn. Results such as A and A′ which are outside the circle but within the parallel lines, show the presence of inaccuracy. Results such as B and B′, lying outside the parallel lines, indicate that there is a major component of imprecision. Matrix effects in one or both specimens may be another explanation. Coding of results on the diagram, an effective display of the relationship of the individual and the method group (say) to overall group performances is readily obtainable. An enlargement of this design, employing replicate analysis of multiple specimens, will permit quantification of both group and individual laboratory performance by the application of regression analysis.

6.5.3. Zender & Linder (1976) (22) also describe a two-sample design in which a laboratory's results x_i and y_i suitably transformed with reference to the ratio $\bar{x}/\bar{y}$, are used to derive an "accuracy" probability score and a "precision" probability score. The probability scores are classified on a six-tiered scale running from >0.5 = "good" to <0.01 = "very poor".

This approach goes further than the Youden plot in seeking to quantify both precision and accuracy from results on a single pair of specimens.

6.5.4. Some authors believe that the most effective way of indicating performance is to use a score derived from each result: e.g. score = (result − target value)/group S.D. Several scoring systems have been developed which gather together individual "performance scores" for all analyses done by a laboratory, and calculate the arithmetic mean, which, over an entire series, will furnish an index of inaccuracy. The arithmetic mean of the modulus (i.e. sign ignored) of this score taken over an entire series will furnish an index of imprecision. This "laboratory performance score" may then be ranked alongside the score of all other participants to encourage continual improvement. One variant of this is to plot the "running mean" score of the last 20 or so tests on a time scale for each participant, to illustrate graphically the individual or group change in performance with time (*Whitehead* et al., 1973) (19). The advantage of such a scoring system ("Variance Index") is that it gives a single index of performance, instead of separate measures of accuracy and precision, and is therefore simpler to use. Disadvantages of these scoring systems are:

(a) It is difficult to relate the score to accuracy and precision and thereby use it to diagnose the cause of poor results.

(b) Several results are necessary to produce a reliable score. Consequently it will take time to produce sufficient results to calculate a separate score for each analyte. Combining the results for different analytes produces a useful overall score of performance quickly, but gives no information about individual analytes.

6.6

After a single distribution of material, parameters of location and dispersion of results for each assayed constituent, and within each defined sub-group, may be made by conventional methods. When the design includes replicate analyses on the same material (Section *5.3.2*), estimates of within-laboratory precision may also be made.

7. Remedial action

The final phase of external QC is the action taken by the participant, which relies initially on a correct interpretation of the type of problem which is being experienced. In many cases, the interpretation itself provides the clue to an appropriate counter-measure. This section attemps to deal primarily with problems revealed by external QC, but obviously these cannot be divorced entirely from those shown by internal QC.

7.1

If a result differs markedly from the target value and has been classified as an outlier,

check that

(a) no clerical error has been made in transcribing the result,

(b) special calculations (e.g. conversion to unfamiliar units, or allowance for dilution) are performed correctly,

(c) the correct specimen was analyzed,

(d) the specimen was not grossly contaminated at the time of analysis. Errors of the type listed below might also give rise to outliers.

7.2

If results are widely scattered on both sides of the target value, but are not outside ± 3 S.D. limits, this indicates imprecision which should be detected and followed by internal QC procedures.

Check that

(a) the analysis technique as laid down is being followed exactly,

(b) staff training, morale, working conditions, etc. are satisfactory,

(c) maintenance procedures, e.g. on pipetting devices, thermometers, are adequate.

If the level of precision attained has never been better than the current level, consideration should be given to replacing either those items of equipment which can be shown independently to give poor precision, or to changing or modifying the method.

7.3

Results deviate consistently from the target value.

7.3.1. Depending on the method for fixing the target value (see Section *6.2*), this may reflect methodological differences only. Even if this is the sole cause, some consideration should be given to a change of technique in an attempt to improve interlaboratory comparability. Obviously, such consideration will include management factors such as time, staff and equipment; and of course, change of method, in an attempt to achieve conformity of results, should not stifle individual endeavour and initiative in seeking to improve the state of the art.

7.3.2. If the target value is derived from the method mean (Section *6.2.5*), more direct action is indicated.

Check

(a) calibration standards which may have been formulated incorrectly, prepared from an impure batch of material, or have deteriorated or evaporated on storage,

(b) pipetting devices as in *7.2* (c). These may have been set incorrectly so that they consistently deliver an incorrect volume. If they have a tendency to "stick", then the incorrect delivery, when it occurs, will always be less than the specified volume.

References

1. Büttner, J., Borth, R., Boutwell, J. H., Broughton, P. M. G. & Bowyer, R. C. (1978), Provisional Recommendation on Quality Control in Clinical Chemistry, Part 5. External Quality Control. Clin. Chim. Acta *83*, 189F–202F; J. Clin. Chem. Clin. Biochem. *16*, 259–266 (1978); Clin. Chem. *24*, 1213–1220 (1978).
2. Büttner, J., Borth, R., Boutwell, J. H., Broughton, P. M. G. (1979a), Approved Recommendation on Quality Control in Clinical Chemistry. Part 1. General Principles and Terminology. Clin. Chim. Acta *98*, F129–F143; J. Clin. Chem. Clin. Biochem. *18*, 69–77 (1980).
3. Büttner, J., Borth, R., Boutwell, J. H., Broughton, P. M. G. & Bowyer, R. C. (1979b), Approved Recommendation on Quality Control in Clinical Chemistry. Part 2. Assessment of Analytical Methods for Routine Use. Clin. Chim. Acta *98*, F145–F162; J. Clin. Chem. Clin. Biochem. *18*, 78–88 (1980).
4. Büttner, J., Borth, R., Boutwell, J. H., Broughton, P. M. G. & Bowyer, R. C. (1980a), Approved Recommendation on Quality Control in Clinical Chemistry. Part 3. Calibration and Control Materials. J. Clin. Chem. Clin. Biochem. *18*, 855–860.
5. Büttner, J., Borth, R., Boutwell, J. H., Broughton, P. M. G. & Bowyer, R. C. (1980b), Approved Recommendations on Quality Control in Clinical Chemistry, Part 6. Quality Requirements from the Point of View of Health Care. J. Clin. Chem. Clin. Biochem. *18*, 861–866; Clin Chim. Acta *106*, 109F–120F.
6. Büttner, J., Borth, R., Broughton, P. M. G. & Bowyer, R. C. (1983), Approved Recommendation 1983 on Quality Control in Clinical Chemistry. Part 4. Internal Quality Control. J. Clin. Chem. Clin. Biochem. *21*, 877–884.
7. Caragher, T. E. & Grannis, G. F. (1977), Design of Quality-Control Specimens for Use with a Small Multichannel Analyzer. Clin. Chem. *23*, 2011–2023.
8. Gilbert, R. K. (1975), Progress and Analytic Goals in Clinical Chemistry, Am. J. Clin. Pathol. *63*, 960–973.
9. Gilbert, R. K. (1976), A Comparison of Participant Mean Values of Duplicate Specimens in the CAP Chemistry Survey Program, Am. J. Clin. Pathol. *66*, Suppl., 184–192.
10. Gilbert, R. K. (1978), Accuracy of Clinical Laboratories. Studied by Comparison with Definitive Methods, Am. J. Clin. Pathol. *70*, Suppl., 450–470.
11. Grannis, G. F., Gruemer, H.-D., Lott, J. A. & McCabe, W. C. (1972), Proficiency Evaluation of Clinical Chemistry Laboratories, Clin. Chem. *18*, 222–236.
12. Jansen, A. P., van Kampen, E. J., Leijnse, B., Meijers, C. A. M. & van Munster, P. J. J. (1977), Experience in The Netherlands with an External Control and Scoring System for Clinical Chemistry Laboratories, Clin. Chim. Acta *74*, 191–201.
13. Pickup, J. F., Jackson, M. J., Price, E. M., Healy, M. J. R. & Brown, S. S. (1975), Use of the Reference Method for Determination of Serum Calcium in a Quality-assurance Survey, Clin. Chem. *21*, 1416–1421.
14. Proceedings International Conference on Standardization of Diagnostic Materials, June 5–8, 1973. US Department of Health, Education and Welfare (October, 1974).
15. Röhle, G., Breuer, H. & Oberhoffer, G. (1974), Collaborative surveys of the German Society for Clinical Chemistry, Mitt. Dtsch. Ges. Klin. Chem., 43–56.
16. Skendzel, L. P. & Youden, W. J. (1969), A Graphic Display of Interlaboratory Test Results, Am. J. Clin. Pathol. *51*, 161–165.
17. Stamm, D. (1971), Ringversuche in der Klinischen Chemie, Schweiz. Med. Wochenschr. *101*, 429–437.
18. Stamm, D. (1974), Introduction of Quality Control for all Quantitative Clinical Chemical Analyses in West Germany, Mitt. Dtsch. Ges. Klin. Chem., 25–32.
19. Whitehead, T. P., Browning, D. M. & Gregory, A. (1973), A Comparative Survey of the Results of Analyses of Blood Serum in Clinical Chemistry Laboratories in the United Kingdom, J. Clin. Pathol. *26*, 435–445.
20. Whitehead, T. P. (1974), Quality Control Techniques in Laboratory Services, Br. Med. Bull. *30*, 237–242.
21. WHO (1981), External Quality Assessment of Health Laboratories, Report on a WHO Working Group, Regional Office for Europe, WHO, Copenhagen.
22. Zender, R. & Linder, A. (1976), Interlaboratory Comparison of both Accuracy and Precision by a Two-sample Method. In Quality Control in Clinical Chemistry (Anido, G. A. et al., ed.), pp. 85–93, de Gruyter, Berlin.

Part 6. Quality Requirements from the Point of View of Health Care

This paper is based on a Provisional Recommendation (*Büttner* et al., 1977, (6)) which has been revised to take account of the comments received, and has been accepted by the Council of the International Federation of Clinical Chemistry by mail ballot in December 1979.

Contents

1. General

1.1

In asking for laboratory measurements, clinicians expect to obtain information on body fluids and tissues of individual patients, which will be relevant for the prevention, diagnosis, or treatment of disease (*Büttner* et al., 1979a, Section 1.1). This general objective covers many widely diverse needs and situations.

1.2

Three consequences of this diversity are relevant to the discussion of quality requirements and should be clearly stated at the outset even though they may be thought to be self-evident.

1) Consumer needs (i. e. health care requirements as determined by benefit to patients, clinical practice, and cost to the community) must be taken into account, in order to avoid wrong management decisions which might result from reliance on internal laboratory criteria alone; for example, an analytical method may be the best available for a given component, but it may still be not good enough for clinical application; conversely, a method may be more sensitive, specific, or costly than justifiable by its use in a particular clinical situation.

2) The discussion cannot be limited to precision and accuracy, or even to the reliability criteria (*Büttner* et al., 1979a, Sections 1.3 and 1.4.3); the very nature of the subject makes it necessary to include factors related to organization, interpretation, and epidemiology.

3) Different purposes (e. g. screening for an abnormality; monitoring a treatment) require different levels of perfor mance. For example, a blood glucose determination may be made to screen for diabetes, to diagnose hyperglycaemic or hypoglycaemic coma, as part of tolerance tests, or for other purposes. The level of performance required in each case may be different, depending on the nature of the clinical question at the time. It is, therefore, unrealistic to stipulate a single set of values of the performance characteristics which would make a method acceptable for all purposes in clinical chemistry (CC).

1.3

A discussion in depth of quality requirements would have to make use of tools such as information and decision theory, signal recognition and cost-effectiveness analysis (see e. g. *Büttner*, 1977). The selected aspects discussed here refer only to the typical problems and situations which are indicated by the section headings. The nature of the subject makes it impractical to consider performance characteristics and other variables separately and in turn.

1.4

Part 6 is not a review or summary of Parts 1 through 5. While familiarity with the topics of Parts 1 through 5 is assumed, Part 6 is an attempt to draw attention to some of the complex relationships that exist between quality requirements and factors involved in performance control.

2. Medical significance

2.1

The potential usefulness of a biochemical variable in the prevention, diagnosis, or treatment of disease should be solidly established by research including clinical investigations, before determination of the variable is offered routinely. It should be noted that, in research, quality control of analytical methods used is just as important for the validity of conclusions as it is in routine applications.

2.2

Information is needed on the following items:

1) The distribution of reference values in healthy subjects (reference interval, "normal range').

2) The influence of the time and the physiological state when the specimen is taken. These involve physiological factors such as age, sex, body mass, menstrual cycle, pregnancy, and external factors such as technique of specimen collection (*Statland* et al., 1974), time of day, food intake, exercise, posture, bed rest; wherever possible, these should be controlled to reduce variation.

3) Description of, and convincing evidence for, the changes ascribable to the disease, treatments, or other clinical states for which the variable is claimed to have medical usefulness.

4) Evidence that biological variation between individuals or within any one individual, as well as analytical imprecision and detection limit, are all small enough to permit the detection of the relevant change in a worthwhile proportion of individual cases (as distinct from a change in group mean values).

2.3

The routine use of laboratory methods without prior demonstration of medical significance on all these counts is wasteful of laboratory resources, and detrimental to the quality of health care.

2.4

The requirement of satisfactory accuracy, which underlies the criteria listed in Section 2.2, is not always met in practice. The state of the art may be such that one or several methods of unsatisfactory accuracy and specificity are in widespread use for the determination of a given analyte. It is then to be expected that results for the concentration of that analyte in a control specimen will depend on the composition of that specimen, and will vary with the analytical method used, and that results from some methods may clearly deviate from the definitive value or reference method value if these exist. In this situation, some clinical chemists see an operational advantage in labelling control specimens with several assigned values ("*Sollwerte*", in German), namely, one for each method, and in using these (rather than one objectively better approximation to the true value) as target values in the control of accuracy.

2.4.1. Whether or not this practice is likely to be adopted widely, it is useful to ponder this approach for other reasons. First, it highlights the ways in which differences between methods can influence the clinical usefulness of a biochemical variable. Secondly, it emphasizes the fact that control specimens and calibration standards have different purposes and require, therefore, different specifications; for a standard, only the best available approximation to the true value is acceptable under any circumstances as the basis for meaningful measurements.

2.5

Quality of performance is anything but static or unchangeable, and as methods are improved through research and technological development, their usefulness is often increased by providing new data on the items discussed in Section 2.2, and by extending medical significance to new areas. For example, improved precision and accuracy will be significant factors in the further development of cluster analysis in CC, i. e. the combined clinical interpretation of a characteristic set of laboratory results for several components determined simultaneously, rather than the customary interpretation of one component at a time. The intuitive evaluation of ratios of two component values (e. g. Na/K, albumin/globulin) is not a reliable substitute for cluster analysis, and their exact evaluation is hampered by the fact that mean, variance, and distribution of ratios are often ill-defined.

2.6

When an apparently obsolete method is about to be replaced, some thought should be given to the possibility that the old method may provide clinically useful information not readily available otherwise. For example, a specific determination of glucose in urine is not a useful substitute for the estimation of "reducing substances" in screening newborns for galactosaemia.

3. Time factor

3.1

A laboratory report is useless to the individual patient if it is too late to influence the immediate medical decision, but delayed reports may be important in other respects such as post-mortem diagnosis, health insurance claims, or research. Clinicians may have an understandable tendency to demand instant laboratory service, while laboratories may prefer to report results according to their own work schedules. In fact, the tolerable delay between the clinician's request and the receipt by him of the report clearly varies with clinical urgency which is, of course, not the same for all situations, and acceptable delays should be specified carefully. Total delay can often be reduced significantly by improving communication, i. e. by minimizing the time used in transit of requests, specimens and reports.

3.2

The laboratory should be organized, through ongoing consultation with the clinical staff, in such a way that results are sent out in time for the intended purpose of the analysis. In practice, and subject to justifiable cost and available resources, laboratories will attempt to have results ready the same day, seven days a week, provided that request and specimen are received before a deadline, usually in the morning.

3.3

Some results will usually take more than a day to be delivered because the analysis takes longer or is not carried out daily. Clinical, laboratory, and above all patients' needs should be carefully examined to detect and resolve potential conflicts. For example, it may make good sense in terms of laboratory organization to carry out a seldom requested analysis only once a week, while from the medical point of view the results of this same analysis may be useless if received more than a day after the request.

3.4

There should be a well-argued procedure for dealing with true emergencies. It is useful to define these at two levels of urgency, e. g. "Immediate" (life in danger), and "Needed Today" (such as serial specimens taken for a functional test after the morning deadline). Laboratory organization must include provision for handling emergency specimens with the minimum of interruption to routine operation. One such provision could be to set up a special unit ("stat" lab.) with separate equipment and manpower, but the most appropriate solution to the problem must evolve in the local situation. In some circumstances (which, of course, should be defined carefully), it may be justified to use a simpler technique, i. e. to sacrifice some specificity, precision, or accuracy for a gain in speed. On request and report forms, such rapid techniques should be identified to prevent unwarranted numerical comparison of results which, because of differences in reference intervals etc., may not be so comparable. One way of flagging these rapid, possibly less reliable, results is simply to state that they are approximate and obtained by an emergency procedure. The assessment and quality control of performance characteristics are, of course, no less mandatory for rapid than for other methods.

4.5

Clinicians should list recently administered drugs on the request forms, so as to enable the laboratory to assess the risk of drug interference with certain analyses, and to inform the clinician accordingly.

4.6

The clinician's contribution to quality control should be to bring up for discussion any unexpected, contradictory, or otherwise problematic results as well as unusual or unacceptable delays in reporting. An agreed procedure should exist for this purpose and should not be allowed to lapse or to be replaced by the doubtful expedient of requesting repeat analyses without prior consultation.

4. Communication

4.1

The quality of laboratory performance in the widest sense, that is, the usefulness of CC in health care, depends to a large extent on the quality of communication between the producers of laboratory information and its consumers. A continuous flow of information must be maintained with a minimum of distortion in both directions.

4.2

A carefully devised fool-proof communication procedure is needed to ensure correct identification of specimens, samples, and results under all circumstances.

4.3

The layout of report forms should be carefully designed for completeness and good readability. In addition to the analytical result, the form should include information to identify: (1) the specimen by name and identity number of the patient, and the date and time of day when the specimen was taken; (2) the type of specimen (e. g. capillary or venous blood; anticoagulant: 24-h urine); (3) the physician or hospital division where the report is to be sent; (4) the type of analysis done; (5) the laboratory where the analysis was done. Both the name of the component measured and the unit used for the result should follow international conventions. Since all this information (except the result itself) should originate from the request accompanying the specimen, time can be saved and transcription errors reduced by reporting results on the request form itself or a copy of it.

4.4

The laboratory should issue instructions on the taking and preservation of specimens; and information needed for correct interpretation of results such as reference intervals and precision data.

5. Testing single specimens for abnormality

5.1

Whether in individual cases or in a screening project, one single specimen per subject is often analyzed to determine whether or not the result is abnormal, i. e. inside or outside the reference interval. This is defined as the interval expected to include a stated percentage (conventionally, 95, 99, or 99,9%) of the values found in healthy subjects. This reference population must be carefully defined with respect to age, sex, or other constraints known to affect the distribution of the variable.

5.2

The correct interpretation of statistical limits should be kept in mind: some values outside the limits may belong to the 5, 1, or 0,1% of normals expected to be outside by definition. These misclassifications of some normal values as abnormal are known as errors of the first kind, and they can occur at the stated and conventionally accepted risk of 1 in 20, 100, or 1000, depending on the probability chosen for the calculation of the limits.

5.3

It seems intuitively obvious that analytical imprecision, by widening both the apparent reference interval and the uncertainty interval of individual results, makes it harder to detect abnormals, and it has been suggested as an arbitrary rule that, for a method to be acceptable for that purpose, its analytical error range should not exceed one-quarter of the "normal range" (*Tonks*, 1968).
If "range" is interpreted as "95% range", mathematical analysis shows that the rule amounts to defining as acceptable a widening of the apparent reference interval by 15% over what it would be if imprecision were negligibly small, i.e. if the reference interval depended on biological variation only. There is no intrinsic logical reason for selecting this particular number.

5.4

Acceptable imprecision could be defined less arbitrarily by relating it to the power function of the statistical test used, i.e. to the frequency of errors of the second kind (in this context, classifying abnormal values wrongly as normal). This approach would require, in addition to the reference interval for healthy subjects, information or assumptions on the distribution of abnormal values, as well as decisions on the acceptable risk of errors of the first and second kind.

5.5

The selection of acceptable risks of misclassification of normals and abnormals, whether in screening patients or for a definition of acceptable imprecision, is a social management decision. The problem of optimization is complex in that at least 8 factors must be taken into account, none of which can be quantified easily. The $8 = 2^3$ factors arise as combinations of 2 types of consequence with 2 types of error and 2 types of recipient, namely, the medical and financial consequences of misclassification errors of the first and second kind on the individual patient and on society. In this approach, each analytical method, or at least several groups of methods, would have to be considered separately: obviously, factors will differ in weight depending on the nature of the information sought. For example, misclassification in the detection of anaemia, infertility, and cancer, respectively, would have very different consequences.

5.6

Borderline results may need special attention. To avoid ambiguous or arbitrary interpretation, it is sometimes useful to define a "doubtful" zone of values intermediary between "normal" and "abnormal". Values in this zone would need confirmation such as further analyses of the same or other specimens, and perhaps additional clinical or laboratory data, before a medical decision could be taken.

6. Changes in individual patients

6.1

The biological variation of mean values between individuals has little influence on the interpretation of results when the patient becomes his own control, that is, when variables are monitored in specimens obtained over hours, days, or weeks, and when changes or trends rather than actual values are the clinically relevant information. This is the case in observing the course of disease, recovery, or physiological change, the effect of therapy, or the response in functional tests. Information on the magnitude and variation of the change to be expected under these various circumstances should be available for proper interpretation of the findings.

6.2

Imprecision as well as any variation in accuracy must be small enough to permit the detection of relevant trends or changes. How small is "small enough"? As pointed out before (Section 1.2 (3)), there can be no general answer to this question: it depends on the circumstances such as magnitude of the change to be detected, number of observations, and physiological variation within the patient. Even a change in, or the absence of, physiological variation (circadian, day-to-day) in a patient can be a sign of disease. Since neither these factors nor imprecision can be manipulated at will, it is up to the laboratory to make clinicians aware of these limitations. In particular, the current magnitude of imprecision should be made known, and it should be stated as a rule of thumb that changes less than twice that figure are often within the laboratory error.

6.3

CC results in a patient's chart should remain comprehensible, valid, and comparable when the patient seeks health care involving reports from several laboratories, for example, when consulting another physician, or after moving to another area. However, if life-long storage of individual patient profiles for future reference is to become more than a futile exercise in filing and retrieval, both the long-term accuracy of methods and the agreement between laboratories will have to be improved considerably; it should be emphasized that the latter objective requires efficient external quality control.

7. Statistical significance

7.1

Scientific methodology requires that the probability of chance occurrence ("statistical significance") of an observation should be assessed before a conclusion is drawn from it and a course of action decided. This sound and undisputed principle is rarely applied with any degree of rigour when laboratory findings in individual patients are interpreted in medical decision-making, and there are indeed various good reasons why it would be unrealistic and counter-productive to advocate and enforce statistical rigour in this situation.

7.2

Statistical tests, by their very nature, are designed to deal with groups rather than individuals, and provide little or weak information on the latter. The values of statistical parameters required in a test (number of observations, degrees of freedom, variances, etc.) are likely to differ for each individual application, and both the choice and the correct use of a statistical method would often (though not always) require more time, manpower and expertise than can reasonably be provided in clinical routine.

7.3

Computer evaluation of individual patient data, whatever its potential, is in its infancy. The instant integration and critical evaluation of many qualitative and quantitative factors with different and changing weights involved in medical decision-making will, for the foreseeable future, rely on medical experience and skill rather than algorithms.

7.4

However, it is part of that skill to use reliable statistical information (say, on imprecision, or on reference intervals) whenever it is available, and it is one of the duties of clinical laboratories to communicate such information to clinicians.

8. Epidemiology

8.1

Despite continuing progress in technology, standardization, and quality control, laboratories continue to differ as well as vary in the accuracy and precision of techniques used. It is very desirable that laboratories should co-operate, or be legally held to co-operate, in reducing the inter-laboratory variation to a point where differences in accuracy would be negligible within a country, or even world-wide.

8.2

Much needed large-scale epidemiological studies (for example, on the effects of climate and nutrition in different populations) can be carried out with the desirable degree of quality only if there is good agreement in results between the collaborating laboratories. The unsolved problem of establishing long-term stability in accuracy and precision stands in the way of extended longitudinal studies, and prevents proper utilization of the mass of information accumulating in the records of CC laboratories all over the world. The quality of research findings and of all medical developments based on them depend to a great extent on the reliability of the methods used in CC. Quality control will continue to provide some of the tools needed to attack and solve these problems.

9. Bibliography

1. Büttner, J. (1977), Die Beurteilung des diagnostischen Wertes klinisch-chemischer Untersuchungen. J. Clin. Chem. Clin. Biochem. *15*, 1–12.
2. Büttner, J., Borth, R., Boutwell, J. H., Broughton, P. M. G. & Bowyer, R. C. (1977), Provisional Recommendation on Quality Control in Clinical Chemistry. Part 6. Quality Requirements from the Point of View of Health Care. Clin. Chim. Acta *74*, F1–F9, J. Clin. Chem. Clin. Biochem. *15*, 95–100 (1977), Clin. Chem. *23*, 1066–1071 (1977).
3. Büttner, J., Borth, R., Boutwell, J. H., Broughton, P. M. G. & Bowyer, R. C. (1978), Provisional Recommendation on Quality Control in Clinical Chemistry. Part 5. External Quality Control. Clin. Chim. Acta *83*, 189F–202F, J. Clin. Chem. Clin. Biochem. *16*, 259–266 (1978), Clin. Chem. *24*, 1213–1220 (1978).
4. Büttner, J., Borth, R., Boutwell, J. H. & Broughton, P. M. G. (1979a), Approved Recommendation on Quality Control in Clinical Chemistry. Part 1. General Principles and Terminology. Clin. Chim. Acta *98*, F129–F143. J. Clin. Chem. Clin. Biochem. *18*, 69–77 (1980).
5. Büttner, J., Borth, R., Boutwell, J. H., Broughton, P. M. G. & Bowyer, R. C. (1979b), Approved Recommendation on Quality Control in Clinical Chemistry. Part 2. Assessment of Analytical Methods for Routine Use. Clin. Chim. Acta *98*, F145–F162, J. Clin. Chem. Clin. Biochem. *18*, 78–88 (1980).
6. Büttner, J., Borth, R., Boutwell, J. H., Broughton, P. M. G. & Bowyer, R. C. (1980a), Approved Recommendation on Quality Control in Clinical Chemistry. Part 3. Calibration and Control Materials. J. Clin. Chem. Clin. Biochem. *18*, 855–860.
7. Büttner, J.,., Borth, R., Broughton, P. M. G. & Bowyer, R. C. (1980b), IFCC Document with a view to an IFCC Recommendation on Quality Control in Clinical Chemistry. Part 4. Internal Quality Control. J. Clin. Chem. Clin. Biochem. *18*, 535–541.
8. Statland, B. E., Bokelund, H. & Winkel, P. (1974), Factors Contributing to Intra-Individual Variation of Serum Constituents: 4. Effects of Posture and Tourniquet Application on Variation of Serum Constituents in Healthy Subjects. Clin. Chem. *20*, 1513–1519.
9. Tonks, D. B. (1968), A Dual Programme of Quality Control for Clinical Chemistry Laboratories, with a Discussion of Allowable Limits of Error. Z. Anal. Chem. *243*, 760–765.

III. Expert Panel* on Enzymes

IFCC Methods for the Measurement of Catalytic Concentration of Enzymes**

* D. W. Moss, London (GBR), Chairman 1978–82
R. Rej, Albany, NY (USA), Chairman
H. U. Bergmeyer, Tutzing (DEU)
G. N. Bowers, Jr., Hartfort, CT (USA), Chairman 1971–75
M. Hørder, Odense (DNK), Chairman 1978–78
M. Mathieu, Lyon (FRA)
L. M. Shaw, Pennsylvania, PA (USA)
J. H. Strømme, Oslo (NOR)

** The Panel produces a series of recommendations, which deal with Part 1. General considerations; Part 2. Method for aspartate aminotransferase; Part 3. Method of alanine aminotransferase; Part 4. Method for γ-glutamyltransferase; Part 5. Method for alkaline phosphatase; Part 6. Reference materials for enzyme measurements; Part 7. Method for creatine kinase.

Part 1. General Considerations Concerning the Determination of the Catalytic Concentration of an Enzyme in the Blood Serum or Plasma of Man

This Recommendation of the International Federation of Clinical Chemistry has been prepared by the Expert Panel on Enzymes and accepted by the Committee on Standards. It has been reviewed by other Expert Panels. This process of consultation has resolved the differences in standpoints on quantities and units in enzymology between the Expert Panel on Enzymes and the Expert Panel on Quantities and Units, referred to in the Provisional Recommendation (1974). This Recommendation embodies the agreed position of the two Panels. It has been approved by the Council of IFCC 1978-09-30 by mail ballot.

Contents

Introduction

The members of the Expert Panel on Enzymes (EPE) of the Committee on Standards (CS) of the International Federation of Clinical Chemistry (IFCC) believe that the selection of a method which is to serve as an internationally agreed method of reference to measure the catalytic activity of an enzyme must rest upon experimental observations which have been verified in many laboratories. These methods are to be called IFCC Methods in the remainder of this paper.

The problem of reaching agreements through the detailed definition of a method is considered here, in contrast to the alternative but complementary route of defining the specifications for reference enzyme materials. Since wide acceptance of IFCC Methods is desired, the goals of this international effort in clinical enzymology must be set out early in the programme for review and comment. Furthermore, the definition of terms used must be clear and unambiguous in order to avoid confusion and misunderstanding. It is the view of the present members of the EPE that the EPE should not, as a group, engage in experimental research. However, unresolved differences of opinion or deficiencies in available experimental data brought to light during discussions within the EPE or between the EPE, its

AMs and the CS should be conveyed to the scientific community so that appropriate research is stimulated, with publication of results in the usual way.

The goals

The central consideration in these efforts to define an IFCC Method should be to achieve a blend of maximum reliability and optimum conditions. Any mechanism which establishes an IFCC Method in clinical enzymology should also include a means for change so that further research is not inhibited.

The diagnostic value of determining the catalytic activity of an enzyme in blood serum or plasma which has made it important in medicine should be maintained or enhanced. Each method should contribute to differentiating healthy persons from individuals with diseases known to be associated with an elevation (or depression) of the catalytic activity.The effect on total activity of alterations in the proportions of multiple molecular forms of a particular enzyme found in human serum must also be considered.

The major aim of an IFCC Method is to achieve maximum reproducibility of analysis both within each laboratory and between laboratories throughout the world. Therefore, the sensitivity, precision, specificity and accuracy of the analytical measuring system must be of primary importance.

It is not anticipated that these methods will immediately receive wide acceptance for use in service laboratories. Therefore, the constraint that they initially be 'routine' methods, compatible with existing mechanized analytical equipment, was dismissed since such a limitation might jeopardize their longterm suitability as IFCC Methods.

The need to reach agreement on and to establish a broad acceptance of IFCC Methods must not impair innovative efforts and systematic research to improve enzyme activity measurements. The scientific demonstration of new conditions which are recognized as significant improvements should lead to adoption of a new modified IFCC Method after suitable review. Once accepted, the newer method becomes the official IFCC Method and the older method is no longer endorsed.

1. Definition of terms

Whenever possible, the recommendations on nomenclature, classification and symbols of the Report of the Commission on Biochemical Nomenclature of the Internation Union of Biochemistry should be followed (1). The following terms need further clarification of their use in this document and subsequent EPE draft proposals or recommendations.

1.1. Nomenclature of IFCC Methods

This must evolve and be consistent with recommendations of the IFCC Reference Materials and Methods Office.

1.2. In the previous sections or those which follow, the term 'optimum conditions' means those conditions supporting maximum catalytic effect at the selected temperature in respect to each of the following:

(a) the identity and concentrations of substrate(s), cofactors, activators, buffer and allosteric factors;

(b) the catalytic concentration of indicator enzyme(s) and of auxiliary enzyme(s) (see 1.5);

(c) the pH and ionic strength of the reaction mixture, and

(d) other variables (e.g. exclusion of known inhibitors).

An optimized method embodies the concept of optimum conditions being modified only to that extent required to achieve maximum reproducibility of the final measuring system. Therefore, a method must first be 'optimized' before it is recommended for consideration as an IFCC Method.

1.3

The term 'serum enzyme' is to be avoided since what is meant is 'an enzyme occurring in serum', e.g. one may say aspartate aminotransferase in serum; but should not say serum aspartate aminotransferase.

1.4

The term 'kinetic method (or technique)' should be avoided when what is meant is the continuous monitoring of multiple data points of the progress of the reaction. Instead of 'kinetic method' the term 'continuous (monitoring) method' is recommended. Ambiguity may result from the fact that kinetic techniques includes 'one point', 'two point' and 'end point' methods of monitoring the progress of a reaction. Kinetic methods are thus also employed to determine substrate concentrations (2).

1.5

The terms to describe the two or more single reaction steps of a coupled test system to measure enzyme activity are: (a) the 'primary reaction', (b) the 'indicator reaction' and, if present, (c) the 'auxiliary reaction(s)'. The source of the enzyme catalyzing the primary reaction is serum or other biological material while the indicator and auxiliary enzyme(s) are incorporated as reagents.

1.6

The terms describing an IFCC Method should not contain eponyms or acronyms other than those accepted by the Commission on Biochemical Nomenclature.

1.7. Catalytic activity (symbol *z*)

Comment. Earlier proposals, such as amount of enzyme, catalytic amount, enzymic activity, and catalytic ability, have been abandoned.

1.8. Definition of kind of quantity

Catalytic activity of an enzyme is a property measured by the catalysed rate of reaction of a specific chemical reaction, produced in a specified assay system.

Comments. Catalytic activity pertains to the enzyme of the original system containing enzyme, not to the reaction mixture. Therefore, catalytic activity is conceptually different from rate of reaction, even if it is measured by that rate of reaction. Catalytic activity is an extensive kind of quantity, i.e. doubling the system containing the enzyme doubles the catalytic activity – this in contrast to an intensive property, e.g. a concentration. The catalysed rate of reaction is the observed rate of reaction minus the spontaneous rate of reaction. Catalytic activity is a derived kind of quantity and is defined as equidimensional with rate of reaction, i.e. with amount of substance divided by time.

1.9. Unit of catalytic activity

Katal (symbol kat)
Alternative unit: mole per second (mol/s)

Note. According to its definition, catalytic activity is measurable in the same SI unit as rate of reaction, i.e. mole per second. However, the special (non-SI) name 'katal' is recommended in the specific context of enzymes and clinical chemistry.

Comments. The advantages of having one name and symbol instead of composites is well known physics. The usefullness of katal becomes evident when further derived units for, e.g., concentrations are considered below. The unit 'katal', as any unit, is independent of assay conditions. In contrast, the measured catalytic activity (of an enzyme in a system) referring to one method is different from that referring to another method. Thus, a reference to the method does not belong with the unit, but with the quantity name. The 'katal', of course, is coherent with SI. The 'enzyme unit' (symbol U) is abandoned; for a given method of assay 1 U = 1 μmol/min = 16,67 nmol/s = 16,67 nkat.

If the method employed precludes the use of this kind of quantity and its units, an arbitrary kind of quantity and unit should be defined.

1.10. Catalytic activity concentration (symbol *b*)

Unit 'katal per litre' (kat/l, kat · l^{-1} or (mol/s)/l, mol · s^{-1} · l^{-1}).

Comments. In actual practice, the name may be shortened to 'catalytic concentration'. The volume refers to that of the original enzyme-containing system, not to that of the assay system. The volume is recommended to be the litre, not the millilitre, 1 U/l = 16,67 nkat/l.

2. Reaction temperature

The set point in degrees Celsius of the reaction temperature is to be assigned as a part of the method. Only one temperature should be endorsed for each method. It is recommended that with as few exceptions as possible, the IFCC Methods be carried out at a uniform temperature of 30,00 °C.

2.1 Temperature control specifications

2.1.1. Accuracy. After initiation of the enzyme catalysed reaction the reaction mixture must be in temperature equilibrium within a range of ± 0,05 °C of the set temperature, e.g., 30,00 ± 0,05 °C.

2.1.2. Precision. The methods, procedures and equipment used for controlling temperature, in mixing specimens with reagents, and in inserting them into the measuring device shall be such that during the entire actual time of observation of the reaction mixture, the recorded temperature of the liquid near its centre varies by less than ± 0,1 °C expressed as the standard deviation of at least 16 measurements of temperature made following exactly the same procedures and using the same reagents as used in determining the catalytic activity.

2.2. Temperature sensor specification

The accuracy and precision of the reaction temperature must be verifiable with an independent temperature-sensing device that is completely separate from the temperature-control system. The temperature-sensing device must have the following characteristics:

2.2.1. Accuracy. The mean value of the reaction temperature shall be traceable to the International Practical Temperature Scale of 1968 (3,4).

2.2.2. Thermal mass, self heat and conductance. The immersion element plus its leads shall have a heat capacity relative to that of the heat capacity of the volume of liquid reaction mixture into which it is immersed of less than 0,01 (1%).

The thermal conductance of the immersion element at the most extreme external environmental temperature difference from that of the liquid in the cuvette, and the self-heat generated, if any, shall be such that the change of temperature caused thereby shall be less than ± 0,05 °C during the specified observation time.

2.2.3. Speed of response. To ensure that significant short-term transient temperature changes are observed, the time constant of the exponential response of the immersion element and its recording system shall be less than 0,1 (10%) of the specified observation time during which the determination of the catalytic activity is made.

3. Method selection

In addition to some of the broad generalizations stated under Goals and in the Definition of terms (Section 1), the following criteria are to be considered:

3.1

A recommended method should be specific enough to be of diagnostic value and this must be taken in account when optimum conditions are evaluated. For example, tosyl arginine methyl ester is hydrolyzed at high rate by thrombin and could thus be considered as an optimum substrate; however, it is also hydrolyzed by many other proteases and is therefore not recommended for a reference assay of thrombin in plasma. Such considerations will apply for the choice of recommended substrates for the acid phosphatase of prostate, lipase, aminopeptidase (arylamidase, EC 3.4.11.2), cholinesterase (EC 3.1.1.8), and other hydrolases.

3.2

Measurement techniques which follow the progress of the reaction directly and continuously are to be preferred over sampling techniques requiring additional manipulations. However, regardless of the technique employed, multiple data points sufficient in number to define clearly the progress curve of each test are essential (2,5,6).

3.3

Calculations of the catalytic activity should be based upon that portion of the progress curve which is zero-order with respect to concentration of substrate(s); i.e. the initial, linear portion of the curve after allowing for mixing effects or lag-phases which cannot be eliminated by changes in the reaction conditions.

3.4

The number and types of operational steps should be reduced whenever possible to avoid excessive errors from multiple pipettings and contact with numerous container surfaces. If possible, combined reagent mixtures should be utilized (7).

3.5

The description of the method should include an Appendix giving the important experimental data needed to establish its blend of reliability and optimum conditions.

3.5.1. In defining a method for measurement of the catalytic concentration of an enzyme which may exist as a mixture of isoenzymes in serum or plasma, the following general problem arises: one cannot predict the relative proportions of the different forms of the enzyme which may occur in any one sample since not only does the severity of the lesion affect the ratio between the isoenzymes, but also different isoenzyme patterns occur in different organs releasing the enzymes.

In specifying conditions for measuring the rate of a reaction which is catalysed by several enzymes or isoenzymes, each with individually distinct catalytic charakteristics, two approaches are available:

(a) The first approach is to select a well-characterized preparation of a single enzyme form and to specify reaction conditions optimal for it. This has the advantage that the appropriateness of the specified conditions can readily be verified with a reproducible source of enzyme. This approach may lose some validity if the reference enzyme is not closely representative in properties of the enzymes of the specimens to which the method is to be applied, since the chosen conditions may be inappropriate to these specimens.

(b) The second approach is to use as the source of enzyme specimens of serum which contain a mixture of enzyme forms approximating those in the specimens to which the method is intended to be applied.

3.5.2. Graphic and tabular format should be used to document the conditions chosen; e.g. plots allowing the evaluation of kinetic constants and results of other relevant studies comparing such factors as alternative buffers and substrates should be given.

3.5.3. The sample volume should constitute a fixed volume fraction of the total reaction mixture. The ratio of sample volume to total volume must be such that any effects due to buffering capacity of the sample are eliminated and the effects of inhibitors, cofactors, metabolites and other enzymes are reduced.

Although the initial rate of reaction theoretically increases with enzyme concentration without limit, providing that the substrate is present in saturating concentrations, in practice a limit is imposed by the finite time required to establish the shape of the progress curve. The design of an IFCC Method should maximize the range of catalytic concentration of an enzyme which can be measured without the need for special treatment of the sample, such as dilution. Since it is impossible to recommend a universally satisfactory dilution procedure for enzyme samples, it should be recognized that, for each method, a range of catalytic activity will exist beyond which the results obtained must be regarded as less reliable, even though a specific sample-dilution procedure may be recommended for use with a particular method.

3.5.4. The establishment of the analytical variability of the method should be made with human serum samples preferably at a catalytic concentration which best separates healthy individuals from those persons known to have catalytic concentrations of the enzyme in serum outside the reference interval due to a concurrent disease process.

3.6. Outline for write-up

In the title give the EC 'recommended name' (formerly the 'trivial' name) and follow with the systematic name and number both in parenthesis. The data should also be easily seen. For example:

IFCC Method (1976) for Aspartate aminotransferase (L-aspartate: 2-oxoglutarate aminotransferase, EC 2.6.1.1.))
Measurement of the catalytic concentration in human serum

Introduction and method selection. Give brief description of the enzyme(s) and any other information which is necessary for clarity.

Principle. Give the reaction formula(e) in such a way as to show the direction of measurement and a brief statement concerning the measuring technique.

Optimal conditions for measurement
Instrumentation and equipment
Reagents
Specimen procurement, transportation and storage
Measurement procedures
Analytical variability
Interpretation
References
Appendices

4. Instrumentation and equipment

The overall performance of the instruments and equipment employed should meet clearly stated specifications to reduce measurement errors from laboratory to laboratory.

4.1

Specifications for the overall spectrometric performance is to be stated for each method. Spectrometric data specifications should include: wavelength accuracy, spectral band width, absorbance range and linearity, drift, and stray radiant energy at the selected wavelength. Photometric errors are often neglected, but can readily introduce systematic relative differences in absorbance of 0,01 to 0,05 (8,9). Errors due to fluorescent effects can be especially troublesome in measurements with reduced NAD. Similar specifications should be given for other types of instruments not involving spectrometry.

4.2

Cuvettes should have parallel faces of optical glass or quartz which transmit light without significantly reducing the light intensity due to absorption, reflections or other scattering effects. The light beam should in no way be influenced by the walls of the cuvette or events on the wall's surfaces, and the internal pathlength should be 10,00 mm and deviate less than ± 0,01 mm (10). If cuvettes with other dimensions are used, the performance specifications must be equal to those above.

4.3

pH meters should be calibrated at the reaction temperature with reference buffers from an authoritative source (e.g. IUPAC or NBS). When the final pH must be adjusted with measurements from a pH meter, specifications to avoid slope errors, sodium-ion errors at high pH and junction potential errors should be given.

4.4

All volumetric glassware should meet stated specifications (BSI or NBS class A or other national equivalents) or should be calibrated by the user. Corrections for cubic expansion of liquids must be made if the glassware were calibrated at a temperature different from the reaction temperature.

4.5

Surfaces in contact with any part of the sample, reagents, or reaction mixture must be chemically clean and free from traces of acids, metals or detergents or other compounds known to interfere with enzyme activity measurements.

5. Reagents

Specifications for reagent chemicals, water and enzymes should refer to an authoritative source. If these are not available or are found to be inadequate to ensure the method's success, the needed additional specifications should be included.

5.1. Reagent chemicals

These reagents must be of a known purity and free of impurities known to influence the reaction rate. For example the need for added specifications to guard against the 'lactate dehydrogenase inhibitor' in reduced NAD used to measure the catalytic concentration of lactate dehydrogenase in serum has been shown (11,12). The following sources have been considered to be 'authoritative'.

5.1.1. Reagent Chemicals, 4th edn. (1968) American Chemical Society, Washington, D.C. (13).

5.1.2. Specifications and Criteria for Biochemical Compounds, 3rd edn. (1973) Publ. 1344, Natl. Acad. Sci.-Natl. Res. Council, Washington, D.C. (14).

5.1.3. Bergmeyer, H. U. (1974) Methods of Enzymatic Analysis, pp. 417–556, Academic Press, New York (15).

5.2. Reagent water

Water specifications must meet the requirements of an authoritative source (16). In addition, known inhibitors of the enzyme in question must, if present, be below the minimum inhibitory concentration. (These minimum levels should be given). Steril water (free of any additives) may be required for reagent solutions which are to be stored for any length of time.

5.3. Reagent enzymes

Enzymes employed as reagents (indicator and auxiliary enzymes of Section 1.5) must be of stated quality (18) and meet specifications of authoritative sources. These reagent enzymes must be essentially free from contamination with other enzymes, metabolites and impurities which might alter the rate of the primary or secondary reaction being followed. If contaminants or side activities are unavoidable, they should be described.

5.4. Comparison material ('standard')

The composition and purity of the comparison material (usually a substrate, product or coenzyme, but not an enzyme material) must receive additional specifications which go beyond Section 5.1. The molar (linear) absorption coefficient (7) of the comparison material must be determined at increasing concentrations under the conditions which prevail in the final reaction mixture.

5.5. Buffers

Buffers should be prepared by using an explicitly stated, easily reproduced technique with clearly defined reagents of specified purity. The buffer capacity should be adequate to hold the pH of the reaction mixture to within preestablished limits. These limits should be such that the change in pH, from any source, during the period of measurement should not introduce relative changes in the measurements of catalytic activity exceeding 0,01 (1 %). Storage time limits for the buffer should be established such that changes in pH or production of rate influencing decomposition products arising from storage will not introduce relative changes in catalytic activity measurements exceeding 0,005 (0,5%). Buffer substances should not react with any compound of the analytical system unless such a reaction is specifically described (e.g. transphosphorylation with some amino alcohol buffers in alkaline phosphatase methods).

6. Specimen procurement, transportation and storage

6.1. Preparation of the individual and obtaining specimen

The catalytic concentration of an enzyme in human serum (or other biological materials) may vary physiologically for a number of reasons. Differences may relate to age, sex, body habits, weight, food or beverage intake, fasting, dietary habits, exercise or diurnal rhythms (e.g. the catalytic concentrations of creatine kinase, aspartate aminotransferase and lactate dehydrogenase in blood serum increase in individuals after heavy muscular exercise). Variations have also been observed from race to race and from occupational, geographic, climatic and other environmental factors. Therefore, even under well-defined conditions of patient preparation and the use of internationally agreed IFCC Methods to measure the catalytic concentration of an enzyme in large groups of the world's population, it will still be necessary in the future to study the individuals within each group in great detail.

6.2. Serum or plasma specimens

If plasma rather than serum is to be used, the type and concentration of anticoagulant employed must be stated and shown not to alter the enzyme activity. Serum or plasma showing visible evidence of haemolysis, lipaemia or excessive colour (pigments) may well interfere with activity measurements and appropriate precautions should be described as a part of the method.

6.3. Specimen stability

Information regarding the stability of enzymes in the freshly drawn blood and separated serum with the passage of time at various storage temperatures should be included as a part of the method. Frozen serum or reconstituted lyophilized enzyme control and calibration materials may show rather extreme variations in activity with time which should be noted (18).

6.4. Other body fluids or tissues – (optional)

Special instructions about patient preparation, details of collection or biopsy procedures as well as handling and storage precautions must be given.

7. Measurement procedure

All manipulatory operations to determine the catalytic concentration of an enzyme should be given in detail.

7.1 Instrumental

Give the nominal wavelength setting (e.g. 339 nm) and the tolerance limits of the band width (e.g. ± 2 nm at half peak height). Explicit details for blank correction, timing of readings and the control of non-enzymatic changes are also required. The temperature control and equilibration for the cuvette, its holder, and transfer of solutions should be also given in details.

7.2. Analytical system

Give these instructions in tabular format as follows:

Pipette	Volume (ml)	Final substance concentrations in reaction mixture
(I) Buffer/reagent mixture	.....	.... mol/l
(II) Serum (Sample)	.....	volume fraction = ...
Incubate at 30,00 °C for *x* minutes		
(III) Initiating reagent	.....	.... mol/l

Mix well and record the changing absorbance at ..... nm for ...... seconds. Calculate $\Delta A/\Delta t$, the average change of absorbance with time (in s^{-1}). Correct for blank reactions.

7.3. Calculation of catalytic concentration

$$b = \frac{V}{\epsilon \times l \times v} \times (\Delta A/\Delta t), \text{katal} \cdot l^{-1}$$

where V is reaction volume (in litre, l, or multiple hereof).
ΔA is absorbance change (in the unit 1),
Δt is the time increment (in seconds, s).
l is pathlength of cuvette (in meter, m, or multiple hereof).
v is serum volume (in litre, l, or multiple hereof).
ϵ is molar (linear) absorption coefficient (in $m^2 \cdot mol^{-1}$, $l \cdot mol^{-1} \cdot m^{-1}$ or multiple hereof).

8. Interpretation

8.1. Variability of the method

The proper evaluation of any draft proposal, recommendation, or final IFCC Method requires a sound statistical-experimental design which ties together the efforts of several laboratories. Each laboratory must receive and handle or collect the specimen(s) in an identical manner if bias is to be reduced. Similarly, replicated analyses on each specimen should be performed according to such a stated design that between-days and within-run variation can be evaluated. The tissue or blood source of any reference enzyme material as well as its solvent matrix must be given in detail (19).

8.2. Observation in health and disease

Data on the catalytic concentration of an enzyme found in body fluids or tissues of apparently healthy individuals and patients are the consequence of the application of a method and, therefore *should not constitute a part of the method* (experience with an IFCC Method can be given in an appendix).

8.3. Sources of error

Give a listing of known drug and other interferences including pharmacologic (in vivo) and measurement (in vitro) effects in the appendix.

9. Appendices

9.1

Appendices should be used to give greater detail concerning background studies, relevant historical information (e.g. the EPE deliberation and instrumental and reagent specification).

10. References

1. Enzyme Nomenclature, Recommendations (1972) of the Commission on Biochemical Nomenclature on the Nomenclature and Classification of Enzymes together with their Units and the Symbols of Enzyme Kinetics. Prepared by the Commission on Biochemical Nomenclature and approved by I.U.P.A.C. and I.U.B., Elsevier, Amsterdam.
   Enzyme Nomenclature, Recommendations (1978) of the Nomenclature Committee of the IUB, published for the IUB by Academic Press, New York · San Francisco · London 1979.
2. Pardue, H. L. (1977) Clin. Chem. *23*, 2189–2201.
3. The International Practical Temperature Scale of 1968 (1969) Metrologia *5*, 35–44.
4. Mangum, B. W. (1974), Clin. Chem. *20*, 670–672.
5. Dixon, M. & Webb, E. C. (1934) Enzymes, pp. 10–11, Academic Press, New York.
6. Moss, D. W. (1972), Clin. Chem. *18*, 1449–1454.
7. Engelhardt, A. & Noetges, A. (1970) Aerztl. Lab. *16*, 42–49.
8. Rand, R. N. (1969) Clin. Chem. *15*, 839–863.
9. Burnett, R. W. (1972) J. Res. Natl. Bur. Stand. Sect. A 76A, No. 5
10. Mavrodineanu, R. & Lazar, J. W. (1973) Clin. Chem. *19*, 1053–1057.
11. McComb, R. B. & Gay, R. J. (1968) Clin. Chem. *14*, 754–763
12. Gerhardt, W., Kofoed, B., Westlund, L. & Pavlu, B. (1974) Scand. J. Clin. Lab. Invest. *33*, Suppl. 139
13. Reagent Chemicals, 4th edn. (1968) American Chemical Society, Washington, D.C.
14. Specifications and Criteria for Biochemical Compounds, 3rd edn. (1972) Natl. Acad. Sci.-Natl. Res. Council Public. 1344, Washington, D.C.
15. Bergmeyer, H. U. (1974) Methods of Enzymatic Analysis, pp. 417–556, Academic Press, New York.
16. Winstead, M. (1967) Reagent Grade Water: How, When and Why? Publication of Am. Soc. Med. Technologists. The Stech. Co., Austin, Texas.
17. Bergmeyer, H. U. (1972) Scand. J. Clin. Lab. Invest. *29*, Suppl. 126, Abstr. 12.2.
18. Brojer, B. & Moss, D. W. (1971) Clin. Chim. Acta *35*, 511–513.
19. Bowers, Jr., G. N., Kelley, M. L. & McComb, R. B. (1967) Clin. Chem. *13*, 595–607.

IV. Expert Panel* on Proteins

Provisional Recommendations

* J. R. Hobbs, London (GBR), Chairman
C. Alper, Boston, MA (USA)
N. Harboe, Copenhagen (DNK)
B. G. Johansson, Lund (SWE)
Th. Peters, Cooperstown, NY (USA), Former member

Provisional Recommendation (1978) on Specification for Human Serum Albumin Standard[1])

This provisional document was not submitted to voting for acceptance by the council of IFCC.

Introduction

There can be no doubt about the need for a standard for human serum albumin, which has become especially necessary with the widespread introduction of immunochemical methods of estimation. Since the major objective of such assays is to find the content of albumin as it occurs in biological fluids, where it seems to exist almost entirely in monomer form, it is essential that the standard should behave in exactly the same way as does the albumin in those fluids. It therefore should have the same binding capacity for the various dyes which are sometimes used in the estimation, and what is most important, it should have the same molecular size, for should it form polymers these will cause errors in several of the immunochemical methods, especially those that rely on migration through agarose. Members of the Panel have examined over 18 commercially available freeze-dried human albumin preparations, and without exception all when reconstituted produce solutions containing 8%–14% of polymer albumin. This has been confirmed elsewhere (1), and because of this problem, together with the errors of reconstitution, the Panel has opted that the human serum albumin preparation should be available as a stable solution.

With the provision of bovine albumin standard which can be used for biuret and *Lowry* methods for estimating the total protein content, (U.S. National Bureau of Standards) the Panel feels that a human albumin preparation which does carry the risk of transmitting hepatitis etc., should not be used for this very commonly required estimation. The specification which follows is therefore primarily intended for methods estimating the actual human albumin contents of biological fluids, e.g. serum, urine, cerebro-spinal fluid, transudates and exudates.

During our studies of available preparations, it was often found that if a solution were made and its ultra violet absorption determined, then the solution was dried to constant weight and again reconstituted to the same original volume, the ultra violet absorption had changed. We wondered whether this was due to changes in lipid content, polymer formation, or even in added preservatives. Since the absorption provides a very useful measure, the Panel believe that once the solution is made it should not be freeze-dried, except to establish its mass content of albumin. It will then be possible to specify the ultra-violet absorption of the solution, and the use of two wavelengths such as 280 nm (2) and 210 nm (3) will act as a further check on purity. A sequence for human albumin has been published (4) and from the amino acid content the relative molecular mass could be calculated as 66458. However, it is usual for natural albumin to bind 1–2 fatty acids per mol, and if these are stripped below 1 per mol the stability of the molecule can suffer. It would seem reasonable at the present time to adopt a relative molecular mass of 67 000 for the proposed standard.

Specification

1. Purpose

International liquid standard for the comparison of albumin assays.

[1]) Stage 2, Draft 1, 1978–10

2. Source

Plasmapheresis of normal persons who have been shown to be negative for hepatitis antigens and antibodies. During the collection of such plasma or blood, the first 20 ml drawn should not be included, because the entry of the needle into the vein releases tissue activators of plasmin, and thereby would result in proteolytic damage continuing to the albumin subsequently fractionated. For the same reasons, the plasma or serum should not be allowed to stand over the blood cells for more than one hour, to avoid the proteases released mainly from white blood cells.

3. Preparation

Albumin can be prepared to about 95% purity by *Cohn* fractionation or by decantation electrophoresis (preferred) and the Panel would discourage the use of acetyl tryptophan, decanol, or other stabilizing or crystallizing agents because of their subsequent interference both with dye-binding and ultraviolet absorption. Stripping of the bound fatty acid is also discouraged. To avoid future polymerisation, the thiol group of human albumin can be blocked using cystine, by the following procedure: *L*-cystine, freshly dissolved in a minimal amount of NaOH (0,1 mol/l is added slowly at 5 mol/mol albumin (10–100 g/l) at pH 7,0–7,5 at room temperature. After five hours some cystine will probably precipitate as a fine white solid and can be removed by filtration or centrifugation. The unbound soluble cystine can be removed by dialysis or deionization or during a subsequent fractionation step by chromatography or gel filtration (5–7). Thereafter the preparation of the pure albumin monomer can be completed by ion-exchange chromatography, gel filtration or other techniques.

4. Criteria for purity

(i) Immunochemical. There should be a single band by electrophoresis conforming to the AA genotype, and by crossed-immunoelectrophoresis (8), performed against an antiserum to whole human serum there should be shown to be no detectable peaks attributable to other proteins. These two methods should detect impurities above 0,5%.

(ii) Molecular sieving. Using polyacrylamide gel, dimer or higher polymers should not exceed 1%.

(iii) Nitrogen content. By *Kjeldahl* this should be at a minimum of 15,7%. The nitrogen content of human albumin has been estimated to be 15,3% (9), 16,1% (10), 16,5% (11) and there is at present no scientific basis for a nitrogen factor of 6,25 to calibrate this protein. Albumin with 2 mol of stearic acid bound has a calculated nitrogen content of 16,45%.

(iv) Sulphated ash no greater than 0,1% w/w.

(v) Hexose no greater than 0,1% w/w.

(vi) Lipid. As fatty acid this should be 1–2 mol/mol, i.e. 0,4–0,9% w/w.

(vii) Hematin absorbance, at 405 nm, of the final product diluted to 10 g/l should not exceed 0,025.

(viii) Non-protein amino compounds (as norleucine) (12) no greater than 0,05% w/w.

(ix) Calcium should not exceed 8,75 μmol/g of protein.

5. Assignment of albumin content

For an albumin solution corresponding to the above criteria for purity, the albumin content could best be estimated by drying a precise volume to a constant weight. Prior to this three alternatives are possible, firstly to dialyse against ammonium formate which should volatilise during the drying; secondly to know the precise content of sodium chloride, and to check this by a conductivity measurement and to subtract the estimated weight of salt; or thirdly to deionise the solution prior to drying provided no albumin is bound to the resin. The residual water content after drying should not exceed 1% w/w and ideally a correction should be made for any detected water.

6. Packaging

By using a solution, of known purity with assigned content, the volume in the sealed non-reusable glass ampoule should be about 1,1 ml. The pH should be between 6,5 and 7,5 and sodium chloride should be added to 0,15 mol/l for stability. The material should be sterile. A suitable concentration of albumin would be from 50 to 70 g/l and the exact value should be stated on the label. The ampoules would preferably be easy to break open and should be stored at 2 °C–6 °C in the dark.

It would be useful to state the absorbance when the solution is diluted to 10 g/l through a 1-cm path at the following wavelengths: 280 (5,8–6,0) (2), 210 (203) (3), 252, and 308 nm. The last two can be used to correct for turbidity in the final solution using the following formula

$$(2 \times A_{280} - 1 \times A_{252} - 1 \times A_{308})$$

On the final label it could merely state A_{280} and A_{210}, these being the absorbances for a 10 g/l solution through a 1-cm path.

7. Methods

Those methods, that have not been included in the list of references, may be found in the Appendix to the Specification ACC 1 Standardised Protein Solution (BSA) 1972, available from NCCLS, 771 East Lancaster Av, Villanova, PA 19085, U.S.A.

References

1. Solli, N. J. & Bertolini, M. J. (1977) Polymer distribution in human serum albumin powders prepared by lyophilization or acetone drying. Vox Sang. *32*, 239–241.
2. Schönenberger, M. (1955) Streulichtmessungen an Plasmaproteinen, Z. Naturforsch. *10b*, 474–478.
3. Tombs, M. P., Souter, F. & Maclagan, N. F. (1959) The spectrophotometric determination of protein at 210 nm. Biochem. J. *73*, 167–171.
4. Meloun, B., Moravek, L. & Kostka, J. (1975) Complete amino acid sequence of human serum albumin, FEBS Lett. *58*, 134–137.
5. King, T. P. & Spencer, M. (1970) Structural studies and organic ligand binding properties of bovine plasma albumin, J. Biol. Chem. *245*, 6134–6148.
6. Isles, T. F. & Jocelyn, P. C. (1963) The reaction of protein thiol groups with some disulphides, Biochem. J. *88*, 84–88.
7. Edwards, E. B., Rombauer, R. B. & Campbell, B. J. (1969) Thiol-disulfide interchange reactions between serum albumin and disulfides, Biochim. Biophys. Acta *194*, 234–245.
8. Wecke, B. (1973) Crossed immunoelectrophoresis, Scand. J. Immunol. Suppl. *1*, 47–56.
9. Sunderman, F. W., Jr., Sunderman, F. W., Falvo, E. A. & Kallick, C. J. (1958) Studies of the serum proteins, II. The nitrogen content of purified serum proteins separated by continuous flow electrophoresis, Am. J. Clin. Pathol. *30*, 112–119.
10. Schultze, H. F., Heimburger, N. & Frank, G. (1962) The amino acid composition of human prealbumin (try) and that of human and bovine serum albumin, Biochem. Z. *336*, 388–396.
11. Peters, T. (1977) Serum albumin: recent progress in the understanding of its structure and biosynthesis, Clin. Chem. *23*, 5–12.
12. Reeder, D. J., Smiegoski, L. T. & Schaffer, R. (1978) *o*-Phthalaldehyde for the fluorometric assay of non-protein amino compounds, Anal. Biochem. *86*, 490–497.

The Human Serum Standard IFCC 74/1[1)]

This provisional document was not submitted to voting for acceptance by the council of IFCC.

Introduction

It is widely recognised that there is a need for a human serum standard for the measurement of albumin, IgG, IgA and IgM and possibly also for transferrin, orosomucoid, alpha$_1$-antitrypsin and C3. With future measurements of these proteins against the standard as the main objective, a programme evolved to prepare a standard of normal human serum, kept under nitrogen at – 70 °C, which after only one thawing remained clear enough to enable nephelometry. The panel together with its associate members also undertook field trials for the practicality of shipping such a standard around the world to enable cross-calibrations to be made against national standards in various laboratories, and from the methodology explored the immunochemical methods were those finally chosen, so that this has now added the further proposal of having reference anti-sera available.

A major fault with one of the pre-existing standards, WHO 67/99 had been that on reconstitution the solution was too turbid for nephelometry. Experience with other serum standards, had shown the instability of any excess lipoproteins which might be contained within the material sampled and the actual process of freeze-drying was known to create polymers in important proteins such as albumin. When reconstitution is made from a freeze-dried standard, the original amount included in the ampoule is critical, as indeed is the reconstitution process and we wished to avoid this source of error if at all possible. For this reason a wet standard was chosen, and its further specification is outlined below.

Specification

1. To avoid differences known to occur with sex, race and age, the pool was to be collected from young male adults aged 20–30 years, who were to be healthy members of the normal Caucasian population of London.

2. Volunteers would be screened for known viruses etc. prior to their main blood donation.

3. Volunteers would also be screened for any marked genetic variation in serum levels of any of the commoner proteins for which the standard might be useful, and at the same time any volunteers showing hyperlipaemia would be excluded.

4. To avoid excess lipoprotein, all the subjects would fast for 18 h prior to the collection of the blood early in the morning.

5. It was known that proteolytic activity could be released during the process of clotting slowly over cells, and to reduce this to a minimum, the first 20 ml of blood that came down the collecting limb was allowed to run away to carry with it most of the tissue plasmin activator released following entry into the subject's vein. Furthermore the main collection was then made into an empty glass blood transfusion bottle to encourage rapid clotting. Pilot studies had shown that less complement conversion occurred when these precautions were taken. It was also considered desirable to remove the

[1)] Stage 2, Draft 1, 1978–10

serum from the clot and cells as soon as possible (within 40 min) to reduce its subsequent contamination with any form of protease from the leucocytes. Other workers had indeed used snake venom extracts to accelerate clotting, but we already knew that these very extracts were rich in proteases and could themselves result in the liberation of further protease from the leucocytes, and we therefore wished to avoid their use.

6. It was to be hoped the subsequent pool would be clear enough to avoid the need for filtration, which could have unknown effects on the representation in the subsequent standard. To avoid any influence from bacterial proteases, the serum immediately removed from the clot would be handled throughout with aseptic precautions.

7. To enable easy subsequent distribution around the world, solid sodium azide was to be added at 1 part in 1000 which we believed enabled the ampoule to travel by air mail at existing temperatures. One ml aliquots (± 0,5%) were to be sealed under nitrogen in ampoules of a kind more convenient to open without spilling the contents, i.e. the type with a constricted neck.

8. The main bulk of the ampoules would be stored at – 70 °C, split between two such deep freezers, so that if there was a major breakdown of one at least half would be safe.

9. Such a standard could be subsequently calibrated by a recommended method for its total protein content.

10. Such a standard could subsequently be calibrated by a recommended method against the best possible human albumin available.

11. It was hoped initially that similar calibrations be carried out against pure representative samples of human IgG, IgA and IgM; however delays for nearly three years finally forced the Expert Panel to resort to calibration against the only existing WHO immunoglobulin standard. For that reason the IFCC 74/1 content of immunoglobulins is given both in I.U. and as the mass concentration, which was considered the best estimate available at that time (see below).

12. It would be possible later to calibrate IFCC 74/1 against purified standard of other human proteins such as transferrin, C3 etc.

13. Stability studies would be undertaken of the standard to see how it behaved under the proposed conditions of storage and shipment.

14. The final proof of the value of any standard is its use in the field, and before it would be considered for release, the Panel felt the standard and recommended methods for cross-calibration should have been tried and practised in routine clinical chemistry laboratories around the world.

15. With such studies came the realisation that reference antisera should be also made available for use in such calibrations.

Preparation of 74/1

Initial samples from 23 Caucasian males, aged 20–30 years, were all kindly tested for Australia antigen and antibody by Professor *A. Zuckermann* of the London School of Hygiene and Tropical Medicine and were all found to be negative. The same samples were also tested by two-dimensional immuno-electrophoresis against a rabbit anti-whole human serum which recognised 42 separate peaks. No genetic abnormalities were detected which would obviously affect the final level in a small pool. There was also no allotypic variation in the electrophoretic mobilities for alpha$_1$-antitrypsin, orosomucoid, C3, transferrin or albumin. None of the chosen subjects showed any hyperlipaemia and all were able to fast for 18 h prior to their main blood donation. All the donations clotted enabling separation of the serum within 40 min, and subsequent analysis could not detect any evidence of complement conversion (less than 4%). After cooling of all the serum donations there was no visible turbidity, and we were grateful to the Wellcome Laboratories of Beckenham, U.K. for their professional facilities for thorough mixing under aseptic conditions with the subsequent ampouling under nitrogen and sealing within glass. These are stored in two separate – 70 °C freezers under the care of Professor *J. R. Hobbs*, Department of Chemical Pathology, Westminster Medical School, 17 Page Street, London SW1P 2AR, U.K.

Calibration of albumin

A provisional specification for a monomer human serum albumin standard had been drafted by the Expert Panel, but had not proved attractive to 5 distinguished manufacturers of human albumin principally on the ground that the cost would not justify the potential returns, either financial or in goodwill, for the relatively small market. The Panel therefore undertook to examine 18 commercially available highly purified fractions of human serum albumin applying the criteria that had previously been drafted. Two materials almost met the specification, but on reconstitution from dry powders produced solutions containing albumin polymers at 12% and 14%, and also showed a variation in the U.V. absorbance as against that specified. At that time, one firm produced a human serum albumin as a solution collected by electrophoresis and this material was virtually all monomer and otherwise met the specification. On drying this to constant weight, after dialysis and conductivity corrections for residual ions, it was possible to assign a mass concentration value to the solution, and it was therefore decided to use this solution as a standard for the calibration of the serum pool, 74/1.

As different methodologies would be used in many laboratories, four common methods were chosen and a collaborative study was undertaken by the Panel together with 9 of its associate members, and subsequently another 5 have confirmed those results. A review on the measurement of albumin in human serum was published (1) from the Chairman's laboratory, and the details of the Panel study are given elsewhere (8). From this it emerged that the method of choice that could be most easily set up and reproduced in laboratories of 12 member nations was the *Laurell* rocket method. A reference rabbit anti-albumin was generously provided by DAKO of Copenhagen, and this yielded a single sharp peak against both the chosen albumin solution, 74/1 and a whole range of human sera both normal and abnormal. In the subsequent two years, together with this original study, the mean value from laboratories in 17 different countries has ranged from only 39,8 to 40,7 g/l as the estimated albumin content of IFCC 74/1 yielding a mean at 40,3 g/l.

These studies used up our supply of the albumin solution chosen, and we were then dismayed to find it was no longer in production. In its place the firm was producing a triply crystallized preparation, which however on reconstitution produced 11% of polymers. The *Laurell* rocket method is susceptible to polymers which result in a lower height for the albumin peak, so that at the present time the Panel is unaware of a human albumin preparation suitable for further calibration studies against any other standard. This experience is confirmed elsewhere (2). The Panel has therefore chosen to adopt as its own human albumin standard the whole serum 74/1 which is at least known to contain 40,3 g/l of human albumin with no polymer within the limit of detection (0,4%).

Calibration for IgG, IgA and IgM

An IUIS sub-committee was exploring the preparation of pure human immunoglobulin standards from 1974, and so through its liason with this group the Expert Panel awaited the possibility of undertaking primary calibrations against IFCC 74/1. In August 1979 it was learned that the immunoglobulin project had been abandoned, and it was therefore decided to calibrate IFCC 74/1 against WHO standard 67/99, which alas was from material collected in 1965. However, in intervening years many workers had become familiar with the international unitage of this WHO standard, and indeed the WHO had published (3) the mg/ml equivalent of those units by the best methods currently available. In clinical usage most of those interpreting immunoglobulin levels seem to prefer to do it in mg/dl or indeed in g/l and so the Panel adopted the latter (conforming to S.I. recommendations) for its own use. It was realised that the g/l equivalents were not absolute, but what is important is that they were internationally acceptable and in agreed use. If at some time in the future better calibrations for pure human immunoglobulins were undertaken, it will be possible to reassign new values for international use.

Each ampoule of WHO 67/99 had an average 102 I.U. assigned to it and in order to avoid any between-ampoule differences it was decided to pool the contents of ten ampoules. By radial immunodiffusion assays the arbitrary mass concentration values had been assigned (3) to 67/99 as IgG 7,58; IgA 1,34; IgM 0,80 g/l. For any comparison with WHO 67/99, which produced a turbid solution on reconstitution, the method of choice had to be radial immunodiffusion, although there was inadequate information about rocket methods or indeed the possibilities for compensated nephelometry. At that time the five Protein Reference Units of England and Wales were all equipped to undertake a study of the methods which indeed showed (4) there was method-specific variation in calibrations between WHO 67/99 and IFCC 74/1. These results therefore forced the choice of the radial immunodiffusion comparisons for the purposes of the calibration. From these it can be stated (the values in brackets are the standard errors of the estimates) that IFCC 74/1 contains IgG at 108 I.U./ml (2,6); IgA 102 I.U./ml (1,6); and IgM 146 I.U./ml (3,0). From these, the arbitrary weight units can be assigned as IgG 8,68; IgA 1,45; IgM 1,24 g/l.

Throughout the above estimations the actual antisera used were from goats and were obtained from Atlantic Antisera Inc. using Batch No. P033N for anti-IgG; Batch No. P034N for anti-IgA; and Batch No. P028N for anti-IgM. In that further supplies of these batches were limited, the same laboratories have undertaken a further comparison of antisera from six different sources from which might be chosen reference antisera to be held on behalf of the Expert Panel, available for cross-calibrations by members of the IFCC[2]).

From the above study (4) it became clear that WHO 67/99 is unsuitable for the estimation of IgM by nephelometry, yet gave a virtually identical result by formylated rocket assay. For IgA the radial immunodiffusion result was lower than that by formylated rocket assay or by nephelometry and the reasons for this are not yet clear. For IgG, radial immunodiffusion gave slightly higher results than the rocket method and slightly lower results than the nephelometric method.

Throughout the comparisons, a computer was used to analyse the raw readings from each of the five laboratories and to check for parallelism between the calibration curves. It was found that the best parallelism was achieved using the radial immunodiffusion method, and it is an interesting observation that when the *Mancini* curves were drawn by computer the coefficients of variation between batches and between laboratories were the lowest we have yet encountered. It could be

[2]) Antisera from Seward Laboratory, UAC House, Blackfriars Rd., London SE 1, U.K. and from Kallestad Laboratories Inc., 1000 Lake Hazelrize Drive, Chaska, MN 55318, U.S.A., were found suitable: each has kindly donated 100 ml for each Ig class.

said that the average laboratory worker who draws the curves by hand achieves a C.V. some 5% higher than that found by computer, which produced C.V. as low as 2,8% which we had never achieved previously. Thus of the current methodologies examined, radial immunodiffusion was the best choice for cross-calibration provided the curves were drawn by computer, and it is therefore recommended for future calibrations between IFCC 74/1 and national standards, with regard to IgG, IgA and IgM.

Total protein

Total protein, though widely used, must always be an approximate measurement because different proteins and mixtures give slightly different colours and absorbances. The Expert Panel has therefore already recommended (5) that the present reference method be that already published (6) using as a reference standard a pure bovine albumin. Armour albumin solution standard is suitable in that bovine albumin almost contains 16% nitrogen, and it was satisfactory in a previous study (6). The Expert Panel had itself already helped with a specification for a standard bovine albumin (7) and this is now available from The National Bureau of Standards of the U.S.A. Using the above reference procedures it is possible to assign a total protein value to any working standard, and it is recommended that a pool of human serum could be so calibrated for local use, thereby avoiding need to use IFCC 74/1 for purposes of calibrating total proteins.

Other serum proteins

There are ongoing programmes to prepare purified protein standards for alpha$_1$-antitrypsin (Sweden), transferrin (Copenhagen and ICSH), orosomucoid (Copenhagen) and the IUIS is preparing a complement standard. As these become available it will be possible to cross-calibrate IFCC 74/1. It would also in the future be possible to include other protein standards that become available.

Stability studies

Ampoules of IFCC 74/1 (containing sodium azide at 1:1000) were stored at – 70 °C, – 20 °C, 4 °C and room temperature (18 °C - 23 °C). When the pool was first ampouled we did not set up accelerated degradation studies at 37 °C and 56 °C.

Quantitation of albumin, IgG, IgA and IgM was made by *Laurell* rocket and nephelometric methods at six month intervals for the next two years and no significant differences were detected between the four lots. These four lots were also compared by two-dimensional immunoelectrophoresis on large plates at 1 and 2 years. Lipoproteins could not really be evaluated by this method, but another 42 peaks were visible and no significant differences were detected in shape or height between the four lots, with the exception of C3.

The C3 results from the material kept at 4 °C and room temperature showed 14%–17% conversion, whereas the frozen samples showed less than 7% conversion at 1 and 2 years. At 3–4 years C3 suddenly showed 30% conversion in both the frozen samples from – 20 °C and – 70 °C and this had risen to 37%–45% conversion for the other two samples. This sudden change does not conform to degradation calculations and its cause is not known. It has very little influence on the total C3 level, but of course invalidates the use of IFCC 74/1 for conversion studies.

After four years the material has a faint opalescence probably due to degeneration or the low amounts of lipoproteins which were initially present, but this is so faint that it does not register in nephelometric methods at the present time.

Recommended uses

To date, the associate members of the following countries have agreed to undertake cross-calibrations between IFCC 74/1 and locally submitted standards using these recommended methods of the Panel and the reference antisera which will be provided: Australia, Austria, Belgium, Canada, Denmark, G.D.R., Egypt, Eire, France, Italy, New Zealand, Nigeria, Norway, Sweden, U.K., U.S.A., F.R.G., Yugoslavia.

The Panel considers that this service would make the best use of the new standard and it is also hoped to operate a quality-control scheme for all the participating associated members, while this remains of reasonable size.

This would not debar making the standard available to commercial companies, and other international bodies, who might indeed prefer to make use of the above cross-calibration quality-controlled scheme based on the tried and recommended methods and antisera.

It would be hoped that as other pure protein standards become available, it would be possible to assign values to IFCC 74/1 together with recommended methodology and where necessary reference antisera.

Finally, for those associate members who are undertaking a national service commitment, it is proposed to send them a small stock of the standard, reference antisera, and details of the recommended methods.

Conclusions

It has been possible to prepare a wet standard of normal human serum proteins with satisfactory stability with the exception of complement and lipoproteins, and with aceeptable mild opalescence. This has avoided denaturation subsequent on lyophilisation and reconstitution problems.

It has been found that a major cause for discrepancies between laboratories has been in the methodologies used. The selection of a recommended method which has actually been put to the test in the proposed field of use, has produced precision between laboratories which we have not ever before achieved. This has also necessitated the provision of adequate reference antisera for use in the immunochemical methods.

We have therefore learned that it is not just enough to provide a standard, or even to try and get at the mass value content for particular proteins, but that it is imperative to have reliable methods and antisera for the subsequent cross-calibrations. The organisation of an Expert Panel with its associate members within the IFCC offers an excellent means of achieving this, and we are most grateful to all those who have co-operated.

For the future, it is clear that stable standards for lipoproteins, for complement components, and for purified human immunoglobulins are going to be very difficult. Those for alpha$_1$-antitrypsin, orosomucoid, and transferrin are almost ready, and it will be possible to relate them to IFCC 74/1.

References

1. Slater, L., Carter, P. M. & Hobbs, J. R. (1975) Measurement of albumin in the sera of patients, Ann. Clin. Biochem. *12*, 33.
2. Solli, N. J. & Bertolini, M. J. (1977) Polymer distribution in human serum albumin powders prepared by lyophilization or acetone drying, Vox Sang. *32*, 239.
3. Rowe, D. S., Grab, B. & Anderson, S. G. (1972) An international reference preparation for human immunoglobulins G, A and M: content of immunoglobulins by weight, Bull. Wld. Hlth. Org. *46*, 67.
4. Whicher, J. T., Hunt, J., Perry, D. F., Hobbs, J. R., Fifield, R., Keyser, J., Kohn, J., Riches, P., Smith, A. M., Thompson, R. A., Milford Ward, A. & White, P. (1978) Method-specific variations in the calibration of a new immunoglobulin standard suitable for use in nephelometric techniques, Clin. Chem. *24*, 531.
5. Newsletter 13 of IFCC (March 1976) pp. 4–5.
6. Doumas, B. T. (1975) Standards for total serum protein assays. A collaborative study, Clin. Chem. *21*, 1159–1166.
7. ACC 1 Standardised Protein Solution (BSA) (1972) Publ. NCCLS, 2525 W 8th St. Los Angeles, CA 90057, U.S.A.
8. Hobbs, J. R., Carter, P. M., Peters, T. & Harboe, N. (1979) The estimation of normal and abnormal levels of human serum albumin by four methods laboratories, to be published.

V. Expert Panel* on Instrumentation**

* R. Haeckel, Hannover (DEU), Chairman
Ch. Collombel, Lyon (FRA)
T. D. Geary, Adelaide (AUS)
F. I. Mitchell, London (GBR), Chairman 1974–79
R. G. Nadeau, New Jersey (USA)
K. Okuda, Osaka (JPN)

** The Panel has produced guidelines for listing the specifications of various instruments.

Guidelines (1984) for Listing Specifications of Spectrometers in Clinical Chemistry[1])

Prepared for publication by
G. Bechtler, R. Haeckel, H. Lappe, W. Tausch †

These Guidelines were prepared for the Expert Panel by G. Bechtler, R. Haeckel, H. Lappe and W. Tausch. After extensive review by the Panel the draft was circulated widely to scientists in clinical laboratories as well as in industry. The proposals have been endorsed by the Area Committee on Instrumentation of the American National Committee for Clinical Laboratory Standards and suggestions have been included from the Expert Panel on Quantities and Units of IFCC. A provisional document was published in J. Clin. Chem. Clin. Biochem. 18 (1980) 445–449, Clin. Chim. Acta 103 (1980) 249F–258F and in Clin. Chem. 27 (1981) 187–191. The Scientific Committee and the Executive Board of IFCC found the Guidelines to be of such a nature as to not require approval by the Council. The project is thus successfully concluded with the publication of this Stage 2 Document.

Contents

1. Introduction

1.1

Spectrometers[2]) available for use in the clinical chemistry laboratory (absorption spectrometers for the visible and near ultraviolet range) vary considerably in their complexity. The Expert Panel considered that a standard format for the specification of these instruments should be prepared in order to ensure maximum understanding of their relative merits and give quick and comprehensive information of practical relevance.

1.2

It is suggested that manufacturers should use the following guidelines in the description of instruments. They should than state that they have done so, but if such a statement is made, no information mentioned in the guidelines must be excluded, unless it obviously does not apply to the instrument in question.

---

[1]) Stage 2, Draft 2, 1981–08. Publication delayed until 1984.
[2]) The term "spectrometer" is recommended by the International Union of Pure and Applied Chemistry (1).

1.3

The terminology applied is sometimes provisional and will be modified later in the light of any relevant international agreement.

1.4

Several characteristics of a spectrometer are interdependent. Therefore, information is required on band width, false radiant energy, imprecision, response time and sample volume for all the specified wavelengths.

1.5

Similar guidelines are at various stages of preparation for automatic analyzers and flame emission and atomic absorption spectrometers.

2. General information

2.1

Date of completion of listing the following specifications by the manufacturer.

2.2

Name and address of manufacturer.

2.3

Model name or disignation (it is desirable that a photograph of the instrument is included).

2.4 Type of spectrometer

Absorption[3]) spectrometers are usually classified in clinical chemistry by the mode of wavelength selection (monochromator or filter). Filter spectrometers can be supplied with a continuous spectrum or with a spectrum-line radiation source, however, it is common practice to use the term filter-photometer for instruments with a continuous spectrum radiation source.

In these guidelines the following classification is used:

(a) spectrometer with a radiation source emitting a continuous or distinct line spectrum and with a monochromator selecting wavelengths of radiation used for sample analysis (commonly used term: spectrophotometer).

(b) spectrometer with a radiant source emitting a continuous or distinct line spectrum and with (a) filter(s) selecting wavelength(s) of radiation used for sample analysis.

(i) filter spectrometer with continuous spectrum radiation source (commonly used term: filter photometer).

(ii) filter spectrometer with spectrum line radiation source (commonly used term: spectrum line photometer).

2.5 Optical configuration

2.5.1. Single beam. Sample and reference are consecutively measured in the same beam (2).

2.5.2. Double beam. Sample and reference are measured in two different beams of the same or different wavelengths (2).

2.6 Usable spectral range

The wavelength range within which spectrometric measurements meet the specifications claimed by the manufacturer. For filter spectrometers, the nominal wavelength of all filters available in this range are to be specified. If a broader range is declared, it must be stated which specifications are not guaranteed by the manufacturer.

2.7

Free choice of analytical procedures?

2.8

Special features of the instrument which are not specified in the guidelines.

3. Source of radiation

The type(s) of radiation source used in a spectrometer should be stated (e. g. tungsten, metallic vapour, deuterium lamp).

4. Selection of wavelength

For absorption spectrometry the radiant energy which passes through the sample must be selected from the radiant energy emitted by the lamp. True monochromatic radiation cannot be achieved in practice and the radiant energy selected consists of a waveband, the narrowness of which is one important criterion for the quality of a spectrometer.

4.1 Mode of wavelength selection (type): monochromator or filter

Both types need further characterisation. A monochromator is a device or an instrument which, with an appropriate energy source, may be used to delimit continuous bands of electromagnetic energy calibrated in terms of wavelength (in nm) or wavenumber (in m^{-1}) within the usable spectral range, by means of gratings or prisms. Filters are designed to transmit energy of a preselected wavelength band.

---

[3]) Absorption spectrometry, employed in the ultraviolet and visible regions of the electromagnetic spectrum, is founded upon the principle that each substance has characteristic properties which determine the wavelengths of radiation it will absorb. That is, each different atomic or molecular structure of a chemical substance has unique resonant frequencies or energy levels. When the frequency of a particular radiation directed into the substances coincides with resonant frequency or energy level, its radiant energy is imparted to the substance with a certain probability. The phenomenon is referred to as absorption.

4.1.1. Monochromator

4.1.1.1. Type (gratings and/or prism)
4.1.1.2. Single or double monochromator

4.1.2. Filters: type

4.2 Inaccuracy of wavelength selection

This is defined as the maximum difference (nm) between the effective wavelength (3) and the nominal wavelengths[4]) specified with a filter or by a monochromator setting (3). See note under Section 5.3.

4.3 Position of wavelength selection

The monochromator or filter can be positioned either before or after the radiation has passed through the sample.

4.4 Wavelength scanning or filter selection

4.4.1. Manual and/or mechanized.
4.4.2. Continuous change of wavelength to obtain an absorption spectrum.

5. Relative instrumental spectral function

The relative instrumental spectral function (also called: detected radiant power spectrum) $\varphi(\lambda)$ is proportional to the product of the relative spectral distribution of the radiant energy, the relative spectral transmittance of all optical parts and the relative spectral sensitivity of the detector in operation (2). Each wavelength selected has its own relative instrumental spectral performance.

5.1 Half band width

The half band width $\Delta\lambda_H$ is the distance between the two wavelengths of the function $\varphi(\lambda)$ at which the value of this function has been reduced to half of its maximum (2). See note under Section 5.3.

5.2 One-hundredth band width

The one-hundredth band width[5]) $\Delta\lambda_h$ is the distance between the two wavelengths of the function $\varphi(\lambda)$ at which the value of this function has been reduced to 0.01 of its maximum (2). See note under Section 5.3.

5.3 False radiant energy ratio (false light)

False radiant energy is commonly defined as that radiant energy coming from the light source and reaching the detector, with wavelengths which do not correspond to the wavelengths selected for measurement purposes[6]) (4).
The false radiant energy ratio is the ratio between the signal originating from false radiant energy and the total signal as measured without absorbing sample at the wavelength setting referred to.

Note: For the purpose of these guidelines the following four wavelengths should be considered for sections 4.2, 5.1, 5.2 and 5.3: 340 (Hg 334, if not available Hg 365) nm, 405 (Hg 405) nm, 550 (Hg 546) nm and 625 (Hg 623) nm.

6. Cuvette and cuvette compartment

6.1 Cuvette

6.1.1. Type (e.g. funnel, suction, flow-through cuvette), fixed or interchangeable.

6.1.2. Material of cuvette if supplied with the instrument: e.g. synthetic or natural quartz glass, special optical glass, polyacryl, polystyrene. State the usable wavelength range in parentheses (nm).

6.1.3. Geometry of cuvette: rectangular or cylindrical (specifying if axis is perpendicular or parallel to light beam), physical pathlength in mm (3). The tolerance (maximum deviation from the mean, or standard deviation) of the pathlength should be declared.

6.1.4. Minimum volume for measurement (μl for a pathlength of e.g. 10 mm, meaning the smallest volume of liquid in the cuvette necessary to produce a measurement which meets the technical specifications claimed in these guidelines by the manufacturer. Volumes required for filling inlet and outlet tubing or for washing the system (to reduce carry-over[7]) effects to less than 1 %) should be declared e.g. for flow-through cuvettes.

[4]) The nominal wavelength is the wavelength displayed on a variable wavelength instrument, or the preset wavelength in the case of a fixed wavelength instrument (3). Effective wavelength refers to a central tendency of the spectral distribution of radiant energy that emerges from a monochromator and to which a detector responds (3).

[5]) The declaration of one-hundreth band width is not required for highquality spectrometers with a monochromator; then the one-hundredth band width approximately has the double value of the half band width. With filter-spectrometers the declaration of the half band width is not sufficient because of different spectral properties of filters. In clinical chemistry an absorbance of 2.0 and higher is often measured, therefore, the spectrometric accuracy can be better judged from the one-hundreth-band width than from the half-band width. It is recommended to declare both the half-band and the one-hundredth-band width for all spectrometers to allow a better comparability.

[6]) More exactly, false radiant energy has been defined as the total radiant energy of wavelength exceeding the 1.01 fold one-hundredth band width (2). In spectrometers using monochromators for spectral selection false radiant energy is often called stray light. This term is not applicable to spectrometers using filters for spectral selection where false light is caused by background transmittance. External radiant energy is excluded from this definition.
The influence of false radiant energy can be declared for practical reasons as relative absorbance error at A = 1,0 for several wavelengths. These values are calculated from the value measured at $\tau = 1$ (A = 0). For a sample with absorbance 1,0 at the nominal wavelength (within the one-hundredth band width times 1,01) and – in the worst case – absorbance 0 at all other wavelengths the relative absorbance error will be 4.34 times the false radiant energy ratio.

[7]) Carry-over determined according to ref. [5].

6.2 Cuvette holder

6.2.1. Temperature control. The cuvette holder must be characterised with regard to its maintenance of constant temperature and to the type of temperature control (e.g. water circulator, heater, peltier) and to its maintenance of constant temperature (temperature range, allowable deviation from temperature set, time required for sample to reach the temperature required). Maintenance of constant temperature can be achieved either internally (integrated into the instrument) or externally (by a thermostat outside the spectrometer).

6.2.2. Number of cuvette positions.

6.2.3. The reproducibility of cuvette positioning may be stated as relative standard deviation measured e.g. with a solution of A = 1,0 in the same cuvette repeatedly replaced into the cuvette holder.

6.2.4. Movable (manually, motor-driven, programmable).

6.2.5. Maximum usable pathlength.

6.3 Temperature measurement

6.3.1. Is the cuvette compartment accessible for direct temperature measurement by the user?

6.3.2. Temperature measurement built into (the wall of) the cuvette?

7. Detector

The type of the detector has to be specified (e. g. photomultiplier, photocell, photo-conductive cell, phototransistor, photodiode).

8. Signal processing and display

8.1 Analogue display (type)

8.2 Digital display

8.2.1. Number of digits (range of indicator).

8.3 Type of scale

8.3.1. Transmittance
(τ: 0 → 1, resp. transmission (τ: 0–100 %))
8.3.2. Absorbance (A: e.g. – 0,3 → 3,0)
8.3.3. Concentration units (with factor and/or standard)
8.3.4. Results of kinetic measurements.

8.4 Zero compensation (absorbance)

8.4.1. Is zero compensation (A = 0) performed either manually or automatically (push button: installed or available as an accessory)?

8.4.2. Range of zero compensation for analogue signal output or readout, indicating up to which absorbance compensation to zero is possible (under the condition that the signal deviates less than A = 0.001, determined as standard deviation with a response time less or equal to 1.0 s. This information is required for several wavelengths according to note under Section 5.3.

8.5 Accessory signal outputs

8.5.1. Voltage output (e. g. for connection to recorder).

8.5.2. Type of interface with electronic data processor if included in the instrument (IEEE 488, RS 232 etc.). If a non-standard interface is applied, a brief description should be included.

8.6 External control. If external control by a computer is possible, the functions of the spectrometer which can be controlled, are to be listed.

9. Reliability of spectrometric measurement

This information is required for several wavelengths according to note under Section 5.3.

9.1 Spectrometric inaccuracy

Spectrometric inaccuracy is defined as apparent absorbance minus true absorbance (3)[8].

9.2 Spectrometric imprecision

The spectrometric imprecision should be determined in a between-days replication design according to ref. (6).[9]

9.3 Stability

The overall stability of the spectrometer should be declared by the maximal deviation of the indicated value at A = 0.1 during 60 min (after the warm-up period, according to section 10.3) with constant environmental (room) temperature (23 °C).

10. Additional technical data

10.1 Bi- or multichromatic measurements

10.2 Description of microprocessor support for instrument and/or data handling

10.3 Warm-up time (minutes)

The warm-up time is the period required to achieve the technical characteristics claimed by the manufacturer (2). Examples of typical warm-up times or other procedures involved (e. g. stand-by) may be stated (2).

10.4

Power supply: voltage (V), frequency (Hz) and power consumption (VA). The interval of line voltage and frequency fluctuations, which can be tolerated to achieve the manufacturer's specification, should be declared. These declarations are not sufficient to cover all interferences which can be caused by electric fluctuations. Any other interfering factors should be mentioned.

---

[8]) The deviation of the apparent value from the true value (Δ A) should be determined at an absorbance of about A = 1,0, A = 2,0 (and A = 3,0 if possible) as the mean of 15 independent measurements of the same solutions on several days. The mean deviation is given as relative absorbance error Δ A/A. The effect of false radiant energy and band width are not included because they depend on sample properties. The "true value" comes from a standard solution of which the concentration has been certified by other high-quality instrumentation or by weighing a high purity standard substance.

[9]) The standard deviations as a measure for the imprecision can be calculated from the data obtained at 9.1.

10.5 Temperature influence

The exact technical properties of a spectrometer are temperature dependent. All technical information is therefore referred to a temperature of 23 °C, which is a frequent ambiant temperature in many laboratories. The temperature range in which the spectrometer (including the cuvette holder) meets the manufacturer's specifications should be stated.

10.6 Dimension and weight (mass, kg)

Length, depth, height and weight of the basic instrument without accessories.

10.7 Space requirement

The bench or floor space required to allow for operating the spectrometer, for ventilation etc., should be stated.

10.8

Requirements for special fittings such as air, vacuum or water and water outlet.

11. Accessories

A list of available accessories should be provided with these guidelines. Some examples are given in 11.1 to 11.4, other items may be listed if available.

11.1

Documentation of results.

11.1.1. Printer (numerical or alpha-numerical, number of characters per row).

11.1.2. Recorder.

11.2

Availability of other measuring principles: e. g. turbidimetry, fluorimetry, nephelometry, reflectometry.

11.3

Mechanised sample transport or transfer.

11.4

Automatic cuvette positioning.

11.5

Set of glass filters or other devices to check the reliability of the spectrometric measurements.

12. Instructions for users

Manufacturers should provide purchasers of spectrometers with:

12.1

A comprehensive manual, e. g. following the recommendations of the National Committee for Clinical Laboratory Standards ASI-1 (7).

12.2

Brief operational instructions which explain in a few sentences how a spectrometric measurement is performed (approx. 1 page).

12.3

A trouble-shooting scheme to assist the user to locate the cause of the more commonly occurring faults and to indicate the appropriate action to be taken.

12.4

Instructions for regular maintenance program (a check list).

12.5

A list of evaluations already published should be available on request.

The following information is only to be supplied by the manufacturer or distributor for each local market.

13. Price

13.1

The basic equipment containing a spectrometer with light source, ready for spectrometric measurements at the four wavelengths mentioned under Section 5.3 if available.

In addition the price of the following items may be of interest:

13.2 Mechanised cuvette positioning

13.3 Filters

13.4 Light source

13.5 Printer

14. Service

14.1 Guarantee

14.1.1. Guarantee period (months).
14.1.2. What is included (e. g. material, repair time)?

14.2 Service availability

Can the manufacturer's service be called only during normal working hours or at any time?

14.3

Service cost for labour (per hour) and the cost for travelling.

14.4 Service contract

If a service contract can be offered the cost per year and the number of visits included should be indicated.

14.5 Service centres

The number and location of all service centres in the country of interest should be stated.

15. References

1. IUPAC Commission on Spectrochemical and Other Optical Procedures for Analyses (1972). Nomenclature, symbols, units and their usage in spectrochemical analysis. Pure Appl. Chem. *30*, 563–695.
2. Deutsche Industrie Norm, DIN 58960 (1978), Photometer for analytical tests. Beuth Verlag, Berlin.
3. TSI-3. Standard for Determining Spectrophotometer Performance Criteria (1976). National Committee for Clinical Laboratory Standards, 771 East Lancaster Avenue, Villanova, PA 19088, U.S.A.
4. Paulson, R. E. (1964). Test Methods in Spectrophotometry. Stray-Light Determination. Appl. Optics *3*, 99–104.
5. Broughton, P. M. G.; Gowenlock, A. H., McCormack, J. J. and Neill, D. W. (1974). A revised scheme for the evaluation of automatic instruments for use in clinical chemistry. Ann. Clin. Biochem. *11*, 207–218.
6. Büttner, J., Borth, R., Boutwell, J. H., Broughton, P. M. G. and Bowyer, R. C. (1979). Approved recommendation on quality control in clinical chemistry. Part 1: General Principles and Terminology, Clin. Chim. Acta *98*, F129–F143. – J. Clin. Chem. Clin. Biochem. *18* (1980) 69–77.
7. ASI-I. Preparation of Manuals for Installation, Operation, and Repair of Laboratory Instruments (1972). National Committee for Clinical Laboratory Standards, 771 East Lancaster Avenue, Villanova, PA 19088, U.S.A.

Guidelines (1984) for Listing Specifications of Clinical Chemical Analysers[1]

Prepared for publication by

K. Okuda

These Guidelines have been prepared for the Expert Panel by *K. Okuda* with the collaboration of many colleagues both in Japan and other countries. After extensive review by the Panel early drafts were circulated widely to scientists in clinical laboratories as well as in industry. Many helpful comments were taken into account in the preparation of further drafts. The proposals have been discussed by the Area Committee on Instrumentation of the American National Committee for Clinical Laboratory Standards and again helpful comment has been taken into account. A Provisional Document was then published in J. Clin. Chem. Clin. Biochem. 18 (1980) 947–951; Clin. Biochem. 13 (1980) 244–248, and in Clin. Chim. Acta 119 (1982) 351F–362F. No further comments were received. These Guidelines are thus identical with the Provisional Guidelines except for the title and this preamble.

The Scientific Committee and the Executive Board of IFCC found the Guidelines to be of such a nature as to not require approval by the Council. The project is thus successfully concluded with the publication of this Stage 2 Document.

Contents

[1]) Stage 2, Draft 2, 1981–08. Publication delayed until 1984.

1. Introduction

1.1 Many chemical analysers are now available for use in the field of clinical chemistry but as their specifications are not listed in a uniform format, it is difficult from study of manufacturers' literature to select the instrument most appropriate for a particular use. The Expert Panel considered that a standard format should be prepared to facilitate understanding of their relative merits.

1.2 It is suggested that manufacturers should use the following guidelines in the description of instruments. They should then state that they have done so, but if such a statement is made, no information mentioned in the guidelines must be excluded unless it obviously does not apply to the instrument in question.

1.3 Many of the items listed are not directly applicable to all instruments and appropriate modification is then suggested. It is hoped that in the interest of uniformity, manufacturers, in addition to providing the information requested, will also adhere to the same presentation format. For brevity in this document, items are given only in the form of headings.

1.4 Similar guidelines have been produced for spectrometers in clinical chemistry (1) and are in preparation for flame emission and atomic absorption spectrometers and nephelometers.

2. General information

2.1 Model name and number

2.2 Name and address of manufacturer

2.3 Date of completion of this listing of specifications by the manufacturer

2.4 Brief history of the need for and development of the analyser

2.5 Function and any special feature of instrument or system (maximum 100 words)

2.6 Summary of general specifications for instrument or system
- 2.6.1 Mode of sampling
- 2.6.2 Transport
- 2.6.3 Chemical analytical processing
- 2.6.4 Temperature control
- 2.6.5 Measuring
- 2.6.6 Signal handling
- 2.6.7 Type and format of data output

3. Sampling

3.1 Nature of specimen
- Urine
- Plasma
- Other

3.2 Specimen pretreatment required or available in instrument
- Column
- Dialysis
- Other

3.3 Specimen container
- 3.3.1 Type (shape and material)
- 3.3.2 Total volume
- 3.3.3 Nonusable volume ("dead volume")

3.4 Operation of the system

3.5 Specimen protection from deterioration (prior to, or after presentation to instrument)
- Changes in temperature (not temperature control)
- Evaporation
- Light
- Gas diffusion
- Other

3.6 Specimen identification system (available or not)?
If available, give the method(s)

3.7 Maximum number of specimens per batch without refeeding

3.8 Interrupt capability to insert "stat" or emergency specimens
If available give degree of interference with normal running

3.9 Specimen and test throughput
- 3.9.1 Fixed or programmable
  Is automatic stop mechanism fitted?
- 3.9.2 Time required for each assay and rate of throughput of specimens
  - 3.9.2.1 Maximum and minimum times and throughput rates
  - 3.9.2.2 Tabulate for all chemical methods the times needed for one, five and one hundred specimens

3.10 Mechanism of specimen transport
- Turntable
- Endless chain drive
- Conveyor belt
- Other

3.11 Sampling mechanism
- 3.11.1 Type
  - Piston
  - Rotating valve
  - Proportioning pump
  - Other
- 3.11.2 Brief description of principle
  Drive mechanism of sampler

3.12 Sample aspiration and dispensing
- 3.12.1 Sample aspiration/dispensing, automatic/manual
- 3.12.2 Volume of sample to be picked up/delivered
- 3.12.3 If adjustable
  - 3.12.3.1 Mechanism
  - 3.12.3.2 Adjustment range
  - 3.12.3.3 Adjustment increments

3.13 Carryover
- 3.13.1 Carryover of analyte and reagent[2)]
- 3.13.2 Means for reduction
  (washing sampling tip, coating etc.)

3.14 Inaccuracy[2)]
Sample/reagent ratio should be stated if relevant

3.15 Imprecision[2)]
Sample/reagent ratio should be stated if relevant

4. Chemical processor

4.1 General approach
- 4.1.1 Continuous flow
- 4.1.2 Discrete
- 4.1.3 Centrifugal
- 4.1.4 Combined and other

4.2 Mode of analytical system
- 4.2.1 Reagent
  - 4.2.1.1 Prepacked reagents
    Available/mandatory from instrument supplier; from other suppliers
    Free choice of reagents including home made
  - 4.2.1.2 Form of reagents
    - Powder
    - Tablet
    - Liquid
    - Other
- 4.2.2 Measurement approach
  - 4.2.2.1 Single point/specify, as in *Pardue* (2)
  - 4.2.2.2 Kinetic assay, variability of technique/specify, as in *Pardue* (2)
  - 4.2.2.3 Combination
- 4.2.3 Programmes[3)]
  - 4.2.3.1 Fixed programmes[3)]

---

2) Method of determination should be immediately evident; full details of technique should be available on request.

3) Where 'programming' of any type is mentioned, details should be given of the type of hardware and/or software used, together with the method of user modification, if this is possible.

4.2.3.1.1 Modifiable?
If yes, give the method(s) e.g. punched or magnetic card by user or manufacturer

4.2.3.2 Freely programmable[3])?
If yes, give the method(s) e.g. punched or magnetic card by user or manufacturer; key board

4.2.3.3 Special programmes[3]) available?

4.2.4 Blank correction (yes/no)?
e.g. reagent blank, kinetic blank
If yes, give the method(s)

4.2.5 Profile (multiple assay) (yes/no)?
If yes, give the method(s)

4.2.5.1 Selectable or fixed

4.2.5.2 Sequential (serial) or simultaneous (parallel)

4.2.6 Standards (calibrators)

4.2.6.1 Number of standards (calibrators)

4.2.6.2 Type of standard used e.g. aqueous or protein containing

4.2.6.3 Automatic computation of concentration relative to standards (calibrators) (yes/no)?

4.2.7 Limitations of the system

4.3 Repertoire of analyte assays
Number of different types of assay that may be performed at the same time on the same instrument

4.3.1 Single/multi test – sequential – parallel

4.3.2 Minimum number of sequential or parallel tests at initial installation

4.3.3 Incremental number of sequential or parallel tests which may be added

4.3.4 Maximum number of sequential or parallel tests

4.3.5 If multitest, can operation continue if one or more test units become inoperative?

4.3.6 Is repertoire fixed by manufacturer or chosen by user?

4.3.7 If repertoire can be changed, what is changeover time for each method?

4.3.8 Itemize methods available. Against each method give within-batch and between-batch imprecision[2])

4.3.9 If fixed chemical methods can be modified or different techniques applied freely by user, how is programme[3]) changed: i.e. keyboard, magnetic or punched cards, tape, knobs, plug-board etc.?

4.4 Reagent dispensing

4.4.1 Mechanism
Piston
Valve
Proportioning pump
Other

4.4.2 Number of dispensing stations per instrument/test

4.4.3 Volumes

4.4.3.1 Fixed

4.4.3.2 Adjustable cf. 3.12.3

4.4.4 Inaccuracy[2]). See 3.14

4.4.5 Imprecision[2]). See 3.15

4.5 Reagent addition and mixing

4.5.1 Precautions, corrosive reagents etc.

4.5.2 Reagent volume required for priming and/or dispensing. Reagent consumption per test.

4.5.3 Mixing mechanism
Mechanical stirring
Gas: jet/bubbling
Rotation
Centrifugal action
Mixing coil
Other

4.5.4 Ability to programme[3])

4.5.5 Separation or isolation
If yes, give the method(s)

4.6 Reaction vessel
Test tube (plastic or glass)
Tubing
U-tube
Plastic bag
Centrifugal rotor
Other

4.7 Means of transport and transfer of reaction mixture

4.7.1 Type
Turntable
Endless chain drive
Conveyor belt
Tubing with induced flow
Centrifugal
Other

4.7.2 Manual/automatic

4.7.3 Disposable/re-usable reaction vessel
If re-usable, washing mechanism and fractional carryover[2])
Washing/drying of vessel
Suction
Decantation
Jet washing (tap/distilled water)
Other

4.8 Time control of reaction period

4.8.1 Minimal lag time in seconds between start of reaction and first possible reading

4.8.2 Range of controllable reaction interval (incubation time)

4.8.3 Range of total reaction time during which measurement can be made

4.8.4 Inaccuracy[2]) and imprecision[2]) of time control

4.9 Temperature control and monitoring system

4.9.1 Heating
Bath-type
Electro thermostatic block
Other

4.9.2 Cooling

4.9.3 Range of temperature setting
– Fixed
– Multipoint
– Continuously variable

4.9.4 Temperature control
– Location (reaction vessel, reagent etc.)
– Method

4.9.5 Monitoring
- Location
- Method

4.9.6 Ambient temperature tolerance
- Minimum
- Maximum

4.9.7 Time required to reach selected temperature for systems specified in 4.9.4

4.9.8 Humidity control; if required (inside instrument)

4.9.9 Evaporation loss/if any

4.10 Calibration and/or procedure for maintaining precision and accuracy of results

4.10.1 Amount of time required

4.10.2 Frequency

5. **Signal measuring**

Follow the related guidelines e.g. absorption photometer (1)

Note – It may not be necessary to provide here, as much detail as is requested in the other guidelines. However, it is intended that the general format should be followed and the main points included.

6. **Signal handling**

Follow the related guidelines e.g. absorption photometer (1)

See note under 5.

7. **Data handling**

7.1 I/O Port specification of output format with an example of data
BCD
Serial ASC II
RS 232
Other

7.2 Microprocessor (type, size and function)
Output available? If so, describe

7.3 Output
- Format, fixed or variable
- Method, printout or display

7.3.1 Contents

7.3.1.1 Test data
e.g. concentration or other

7.3.1.2 Specimen identification

7.3.1.3 Additional information
- reference values
- other

7.3.1.4 Units

7.4 Automatic warning
Number and type of errors which are indicated automatically e.g. outside limits, non-linear, abnormal, invalid, other

7.5 Data storage
If yes, give type and quantity of information stored

8. **Additional technical data**

8.1 Requirements

8.1.1 Line voltage, amperage and frequency requirement including tolerated intervals, starting currents, load and continuous load, grounding (3)

8.1.1.1 Power surge protection
Safety shut-off switch

8.1.2 Water supply

8.1.2.1 Tap water, distilled

8.1.2.2 Volume rate

8.1.2.3 Pressure

8.1.3 Air pressure and/or vacuum

8.1.4 Gas

8.1.4.1 Type and purity

8.1.4.2 Volume rate

8.1.4.3 Pressure

8.1.5 Any other supply required

8.1.6 Laboratory environment

8.1.6.1 Humidity

8.1.6.2 Other than temperature

8.1.7 Dimensions (height, length, width)
Mass (weight)

8.2 Miscellaneous

8.2.1 Reagent supply for analyser
If special reagents are required give details of availability in relevant countries

8.2.2 Alarm (yes/no)
Give details of function of monitoring system

8.2.3 Starting period, up to time when first specimen may be inserted

8.2.3.1 From off

8.2.3.2 From stand-by mode

8.2.4 Environmental impact and safety
Corrosive reagents
Toxic waste
Noise
Precautions for operator
e.g. dangerous radiation (laser, UV, γ-ray), vibration, fumes, biological hazards
Other

8.2.5 Maintenance – routine and preventive:
time and frequency
cf. ASI-1 document of NCCLS (4)

8.2.6 Waste handling
e.g. drainage required:
any special requirements for waste disposal

8.2.7 Optional accessories, e.g. recorder

8.2.8 Operator training requirement
cf. TSI-6 document of NCCLS (5)

9. **Information which should be available separately and locally**

9.1 Cost

9.1.1 Instrument

9.1.2 Parts

9.1.3 Options

9.1.4 Training

9.1.5 Single use items:
Cups
Cuvettes
Reagents

9.1.6 Estimated cost per test

9.2 Service and maintenance; degree of maintenance possible with normal laboratory staff; availability of service, e.g. 24h/day and from what location

9.3 Other special information

9.4 A photograph of an outside view of the instrument should be provided together with a block diagram illustrating the method of operation of the main system. If similar diagrams of subsystems would aid clarity, these should be provided also.

10. References

1. Haeckel, R., Collombel, Ch., Geary, T. D., Mitchell, F. L., Nadeau, R. G. & Okuda, K. (1980), Provisional Guidelines (1979) for Listing Specifications of Spectrometers in Clinical Chemistry. Clin. Chim. Acta *103*, 249F–258F; J. Clin. Chem. Clin. Biochem. *18*, 445–449 (1980).
2. Pardue, H. (1977), Clin. Chem. *23*, 2189–2201.
3. TS1-5, Power requirements for clinical laboratory instruments and for laboratory power lines. National Committee for Clinical Laboratory Standards, 771 East Lancaster Avenue, Villanova PA 19088, U.S.A.
4. ASI-1 Preparation of manuals for installation, operation and repair of laboratory instruments. National Committee for Clinical Laboratory Standards, 771 East Lancaster Avenue, Villanova, PA 19088, U.S.A.
5. TSI-6, Guidelines for service of clinical laboratory instruments. National Committee for Clinical Laboratory Standards, 771 East Lancaster Avenue, Villanova, PA 19088, U.S.A.

VI. Expert Group* on Diagnostic Kits and Reagents

Revised Recommendation (1983) on Evaluation of Diagnostic Kits

* M. Rubin, Washington, DC (USA), Chairman
R. N. Barnett, Norwalk, CT (USA)
D. Bayse, Atlanta, GA (USA)
E. Beutler, Duarte, CA (USA)
S. S. Brown, Harrow (GBR)
J. E. Logan, Ontario (CAN)
C. B. Reimer, Atlanta, GA (USA)
J. O. Westgard, Madison, WI (USA)
P. Wilding, Wayne, PA (USA)

Part 1. Recommendations for Specifications on Labelling of Clinical Laboratory Materials[1])

Prepared for publication by J. E. Logan

This paper forms part of a series of recommendations on the evaluation of clinical chemistry kit methods. Others deal with:

Part 2. Guidelines for the Evaluation of Clinical Chemistry Kits (1), (pp. 119–122).

Category I. Provisional Recommendation (1982) on Evaluation of Diagnostic Reagent Sets Used in Health Laboratories with Limited Resources
(LAB 83/1).

Part 1. Selection and Evaluation Using Reference Materials – General Considerations.

Part 2. Evaluation of Performance Using Reference Materials of Analytes Commonly Determined in Blood Serum or Plasma.

Category II. Provisional Recommendation (1982) on Evaluation of Diagnostic Materials and Methods by Comparison with a Reference Method.

Part 1. General Considerations (LAB 83/2-6).

Part 2. Hemoglobin (in preparation).

This Document is a modified version of a Stage 2, Draft 1 Document (2) published also by the World Health Organization (3).

Contents

1. Introduction

The twenty-seventh World Health Assembly urged Member States to take steps as rapidly as possible to control the quality of distributed diagnostic materials. Comprehensive labelling forms a vital part of any such control system. It was subsequently decided that WHO promote the adoption of international specifications for adequate and meaningful labelling of diagnostic reagents. It is hoped that Member States will use these recommended specifications in control programs through their adoption as national regulations.

The Expert Group on Diagnostic Kits and Reagents was established by the International Federation of Clinical Chemistry to respond to the World Health Organization's request in the formulation of appropriate specifications.

[1]) Stage 2, Draft 3, 1983-09

2. Definitions

2.1 Labelling

All written, printed, graphic or other matter on or accompanying clinical laboratory materials. Labelling includes both labels and package inserts.

2.2 Label

Written, printed or graphic matter affixed to or appearing upon the immediate, intermediate or outer container of a clinical laboratory material for identification or description but excluding shipping instructions.

2.3 Package insert or brochure

Written, printed, graphic or audiovisual matter accompanying the clinical laboratory material, but not permanently attached to any part of the package.

2.4 Clinical laboratory materials

Substances, chemical and biological, which are intended by the manufacturer or supplier as an aid in the laboratory investigation of disease or determination of the state of health or for laboratory control of patient management and which are not administered to the patient in order to obtain their intended purpose.

2.5 General laboratory materials

Substances, chemical and biological, which have multiple purposes in laboratories and do not have designated uses.

2.6 Kit

Two or more different clinical or general laboratory materials (excluding reconstituting materials), with or without other components packaged together and designed for the performance of a procedure for which directions are supplied with the package. Custom-made and investigational materials are not included.

2.6.1. Scales for the presentation of test results

The tests may be grouped in accordance with the type of scale for presentation of the result, i.e. binary, ternary, quaternary, quinary or more graduated scale. The common designation for tests using the first four types of scales is coarse-graduated scale tests or coarse-scale tests. The principle of reading the test may be included in the designation e.g. reflectometric coarse-scale test or visual coarse-scale test. These designations bear more information than the term "qualitative" which is not recommended. The designation fine-graduated scale test or fine-graduated test should be used instead of "quantitative test" because this term does not separate these two groups of tests as all tests are quantitative in their nature.

2.7 Reference materials

Chemical biological or other materials which are intended to aid in the establishment of the accuracy of laboratory procedures or to check the accuracy of values assigned to clinical laboratory materials, calibrators or control materials. They may be certified by certain agencies or left in unqualified terms.

2.8 Calibration standards or calibrators

Materials which are intended to be used as an aid in adjusting or standardizing clinical laboratory procedures or instruments. Certain calibrators are frequently called standards in clinical laboratory practice.

2.9 Control materials

Materials which are intended to be used in clinical laboratory procedures as an aid in assessing the imprecision, inaccuracy and other aspects of such procedures.

2.10 Custom-made laboratory materials

Materials prepared by a manufacturer at the request of an individual, collaborating group or laboratory and not available for general distribution or sale. These are excluded from the subsequent labelling recommendations.

2.11 Investigational laboratory materials

Materials which are in the process of development. These are clearly labelled "For investigational use only. The performance characteristics of this product have not been established". They are excluded from the subsequent labelling recommendations.

3. Labels

3.1

The following specifications apply to the labels of clinical laboratory materials, kits and kit components, reference materials, calibrators, control materials, and where applicable, to general laboratory materials. The label on the immediate container provides the information given below (points *3.1.1–3.1.9*). If the container is too small to accomodate all the label information, or if the label would interfere with the reading of test results, the manufacturer *should* provide as much information as feasible on the immediate container label. If all the information cannot be listed on the immediate container label, all the information *must* be given on an outer container (carton or overwrap) or on a package insert if more appropriate. As a minimum the product name, the supplier, the lot or control number and applicable cautionary statements are required to be given on the immediate container. If necessary, abbreviations may be used to conserve space.

3.1.1. Product name. A non-proprietary (or descriptive) name is required. A proprietary name is optional. For a product derived from biological material the generic source should be given where relevant to the proper performance of the procedure.

3.1.2. Quantity or amount. Mass, volume, concentration, numerical count or other applicable quantity designation. The use of the Système International d'Unités (SI) is recommended. In some cases, the number of tests that can be performed or the volume of a reconstituted material may be appropriate.

3.1.3. Lot or control number. An alphabetic and/or numeric designation by means of which the complete manufacturing history of the product can be traced.

3.1.4. Storage information. Those storage conditions necessary to protect the stability of the product must be clearly given. Examples are:

2 to 30 °C (or 16 °C ± 14)
2 to 10 °C (or 6 °C ± 4)
−18 °C or below
−40 °C or below
protect from freezing

Other conditions that may affect stability (e.g. light or humidity) *should* be stated. If storage conditions for the opened or reconstituted product are different from unopened, those conditions *should* also be given.

3.1.5. Expiration date. An expiration date based upon the stated storage instructions if applicable. If the stability period for the opened or reconstituted product is different from that of the unopened product, this *should* be stated. As an alternative to giving an expiration date, the manufacturer *may* choose to direct the user to a procedure given in the insert which may be used to determine the suitability of the product for use.

3.1.6. Supplier. Name and address of the manufacturer, packer or distributor. If this is not the address to which the user may direct inquiries, such information must be provided in the package insert.

3.1.7. Cautionary statement. (If a product can be confused with a product administered to a human such as a drug or a biological preparation.) A cautionary statement such as "for laboratory use only" must be used. If a hazard is associated with the product, appropriate signal words such as: CAUTION: RADIOACTIVE, or internationally adopted symbols should be used.

3.1.8. Intended use(s). The intended use of the product must be stated, possibly by the product name, e.g., polyvalent salmonella, slide agglutination antiserum. This paragraph does not apply to "general laboratory materials".

3.1.9. Potency and purity. Where necessary for the proper performance of the procedure, the potency of titre and the purity of the product *should* be given in IFCC approved nomenclature (4); if other terms are used, they should be explained clearly in the package insert (4.1.6).

3.2

The kit container label lists all the information outlined in Section *3.1* above. Additionally it lists the following:

3.2.1. Kit lot or control number. If kit components are not interchangeable between batches, a common kit lot or control number must be applied to each component and to the kit container. If appropriate, other suitable uniform identification may be used.

3.2.2. Expiration date. The kit container must provide the expiration date of the component in the set with the earliest expiration date.

3.2.3. List of contents.

4. Package insert or brochure

All essential information must be provided in the package insert or brochure in addition to that included on the label.

4.1 Clinical laboratory materials and kits

4.1.1. Product name. A non-proprietary (or descriptive) name, is required. A proprietary name is optional. For a product derived from biological material the generic source or microbial group, type and strain *should be given where relevant to the proper performance of the procedure.*

4.1.2. Intended use(s). The intended use(s) should be clearly specified. As an example: for the (qualitative or quantitative) determination of (glucose) in (serum). Ambiguity should be avoided when a material may have several uses, each requiring different concentrations or preparations, e.g. bacitrin dishes for testing bacterial susceptibility and for presumptive identification of group A *Streptococci.*

4.1.3. Principle of test. A concise explanation of the chemical reaction, techniques, and appropriate literature reference should be provided.

4.1.4. Precautions. A description of known hazards and necessary precautions must be provided.

4.1.5. Suitability for use. If appropriate and available, provide physical, biological or chemical indications of instability or deterioration. Provide descriptions of appropriate tests that the user may apply to assure satisfactory performance, where applicable.

4.1.6. Reagents

a. For those materials supplied by the manufacturers:

(i) State the name, quantity, proportion or concentration of each reactive ingredient for products derived from a biological material or materials. State also, when necessary for the proper performance of the procedure, the generic source or sources, potency, specificity, avidity, sensitivity, or other specifications as appropriate. A statement that the material has been calibrated against a recognized reference material should be included, if applicable.

(ii) The presence and change of any catalytic or nonreactive ingredient such as a buffer, preservative or stabilizer which, if not known to the user, could affect the proper performance of the test or influence the results, must be stated. Documentation confirming the performance of the test subsequent to change in ingredients should be available on request.

(iii) For kits, a list of materials supplied by the manufacturer should be given. If some components are interchangeable with another kit, the manufacturer should identify this characteristic of the kit.

b. Materials not supplied by the manufacturer but required for the performance of the test:

(i) These should be listed and description of the purity, procedures for dilution or mixing and other pertinent information should be given.

4.1.7. Equipment. Any special requirements regarding equipment should be described so that the user may select that most appropriate.

4.1.8. Specimen. Where applicable, a description of:

a. Special precautions regarding specimen collection including special preparation of the patient as it bears on the validity of the test.

b. Additive (e.g. anticoagulants, preservatives) necessary to maintain the integrity of the specimen.

c. Known interfering substances.

d. Recommended storage, and handling or shipping instructions for the protection and maintenance of stability of the specimen.

4.1.9. Procedure. A detailed description of each step required for the proper performance of the test. Points to be considered include:

a. A list of precise experimental conditions which must be met, e.g., pH, temperature, times for specific steps, wavelengths, stability of final reaction mixture.

b. Dilution, reconstitution, mixing or other steps necessary for those reagents supplied by the manufacturer including storage conditions and stability of reconstituted materials. The exact kind of fluid that should be used for dilution or reconstitution should be specified.

c. Calibration procedure including selection and preparation of calibrators, use of blanks, preparation of a standard curve.

d. User quality control procedure including recommended use of appropriate controls.

e. Procedural techniques which may assist the user to perform the test more effectively.

f. Special processing or handling necessary for specimens that may contain known interfering substances.

g. Calculation of results including an explanation of each factor and step in the calculation. If appropriate, a sample calculation may be given.

4.1.10. Specifications and special features. A discussion of the advantages, limitations, accuracy, bias, etc., of the method as may be appropriate. Points to be considered include:

a. Statements of accuracy (as compared to a recognized reference method, if one is available, or to the method used in evaluating the product), precision, specificity, test range, linearity, and sensitivity of the product.

b. Expected results and details of how the data were derived including the population(s) studied.

c. Limitations of the procedure (e.g., possibility of cross-agglutination and its significance).

d. Suggested confirmatory tests as appropriate.

4.1.11. Supplier. Name and address of the manufacturer, packer or distributor. This is the address from which a user can obtain further information concerning the product.

4.1.12. Date. The month and year when the package insert was issued.

4.2 Reference materials, calibrators and control materials

4.2.1. Product name. A non-proprietary (or descriptive) name is required. A proprietary name is optional. For a product derived from biological material the generic source or microbial group, type and strain should be given where relevant to the proper performance of the procedure.

4.2.2. Intended use(s). As an example: (calibrator) for (glucose), determination in (serum).

4.2.3. Composition and source(s). Composition and/or biological source(s) of the analyte or the reactive substance and of the matrix, as appropriate. For reference materials, any additives such as preservatives and stabilizers should be listed and described. For calibrators and control materials, additives should be listed and described to the extent necessary for proper usage.

4.2.4. Value assignment. List:

a. Analyte or reactive substance, value and units in common or usual terms. The relevant property (quantity) should be given in the style recommended by IFCC (4). The use of Système International d'Unités (SI) is recommended.

b. Method for value assignment (by literature reference if appropriate).

c. Reference material(s) used for value assignment.

d. Confidence limits for assigned value(s). For reference materials, the protocol for calculating confidence limits should be described.

e. For materials used in microbiological kits and procedures, any microorganism(s) provided should be identified.

4.2.5. Precautions. A descriptions of known hazards and necessary precautions to be taken.

4.2.6. Suitability for use. If appropriate and available, provide physical, biological or chemical indications of instability or deterioration. Provide descriptions of appropriate tests that the user may apply to assure satisfactory performance.

4.2.7. Preparation for use. A description of how to treat the material for subsequent use. Include stability data and recommended storage conditions for the reconstituted material.

4.2.8. Supplier. Name and address of the manufacturer, packer or distributor. This is the address from which a user can obtain further information concerning the product.

4.2.9. Date. The month and year when the package insert was issued.

5. References

1. Rubin, M., Barnett, R. N., Bayse, D., Beutler, E., Brown, S. S., Logan, J. E., Reimer, C. B., Westgard, J. O. & Wilding, P. Revised Recommendation (1983) on evaluation of diagnostic kits.
   Part 2. Guidelines for the evaluation of clinical chemistry kits. J. Clin. Chem. Clin. Biochem. *21*, 899–902.
2. Rubin, M., Barnett, R. N., Bayse, D., Beutler, E., Brown, S. S., Logan, J. E., Reimer, C. B., Westgard, J. O. & Wilding, P. Provisional Recommendation (1978) on evaluation of diagnostic kits.
   Part 1. Recommendations for specifications on labelling of clinical laboratory materials. Clin. Chim. Acta (1979) *95*, F153–F162, J. Clin. Chem. Clin. Biochem. (1979) *17*, 657–661, Clin. Chem. (1979) *25*, 1499–1502.
3. Guiding principles and recommendations on labelling of clinical laboratory materials: a WHO Memorandum. Bull. WHO (1978) *56*, 881–885.
4. Dybkaer, R. Approved Recommendation (1978). Quantities and units in clinical chemistry. Clin. Chim. Acta (1979) *96*, F155–F183, J. Clin. Chem. Clin. Biochem. (1979) *17*, 807–821.

Part 2. Guidelines for the Evaluation of Clinical Chemistry Kits[1])

Prepared for publication by J. E. Logan

This paper forms part of a series of recommendations on the evaluation of clinical chemistry kit methods. Others deal with:

Part 1. Recommendations for Specifications on Labelling of Clinical Laboratory Materials (pp. 114–118).

Category I. Provisional Recommendation (1982) on Evaluation of Diagnostic Reagent Sets Used in Health Laboratories with Limited Resources (LAB 83/1).

Part 1. Selection and Evaluation Using Reference Materials – General Considerations.

Part 2. Evaluation of Performance Using Reference Materials of Analytes Commonly Determined in Blood Serum or Plasma.

Category II. Provisional Recommendation (1982) on Evaluation of Diagnostic Materials and Methods by Comparison with a Reference Method.

Part 1. General Considerations (LAB 83/2-6).

Part 2. Hemoglobin (in preparation).

It is a modified version of a Stage 2, Draft 1 Document (1)

Contents

1. Introduction

It is difficult to provide a testing protocol applicable to all kit methods. This is because the characteristics of the methods differ widely when one considers the variety of kits in clinical chemistry. For a testing protocol to be efficient, the most critical characteristics of the method of interest should be selected for testing. In general, these are performance or reliability characteristics, such as precision, accuracy, chemical sensitivity, and chemical specificity. It is necessary to consider the particular determination of interest in order to develop a detailed experimental procedure because these characteristics can not always be tested in the same manner for all kit methods.

In developing the testing approach, it should be recognized that it is the performance of the kit method that is to be studied. The testing protocol may employ experiments such as replication, recovery, addition of interfering materials, and comparison of values with those obtained by a definitive or reference method. The protocol should also consider matters like safety, ease of handling, costs, etc.

In cases where reference methodology is not available or seems unlikely to be developed, clinical data are necessary for validation of the performance of the kit method. Relevant clinical data must be made available by the manufacturer, and when necessary, should be independently obtained by the testing laboratory. Such information must clearly define the clinical condition for which the diagnostic kit is reputed to yield useful information and must give the range and confidence estimate of appropriate reference values and expected analytical values associated with the defined clinical condition.

1.1 Scope

This document briefly outlines some general guidelines for the evaluation of quantitative methods. Future documents will specify the procedures applicable to the area of clinical chemistry kit methods. The document is limited to an approach for testing the analytical performance of a method.

2. General guidelines for development of a protocol for testing the analytical performance of a method

The general approach is consistent with the guidelines published by the IFCC Expert Panel on Nomenclature and Principles of Quality Control in Clinical Chemistry (3). However, greater emphasis is placed on the details of the experimental procedure and how the results are interpreted as demonstrating acceptable or unacceptable performance.

2.1 Characteristics of interest

For the analyte of interest, it is first necessary to define the characteristics that are important if the kit method is to be acceptable. These characteristics can generally be divided into two classes: (a) practicability characteristics, such as speed, cost, technical skill requirements, resource requirements, service availability, safety, etc., and (b) reliability characteristics,

[1]) Stage 2, Draft 3, 1983-09

such as chemical specificity, chemical sensitivity, precision, and accuracy. The selection of methods which are suitable for testing should be based on consideration of the practicability characteristics. The testing protocol is primarily aimed at evaluating the reliability characteristics.

2.2 Review of product labelling

Product literature and labels should be reviewed to assess both the adequacy of the information provided and to assist in the selection of methods which are appropriate for the needs of the laboratory. Information provided should conform with the recommendations published by WHO (4) and IFCC (2).

2.3 Experimental studies

The experiments should be designed to estimate the magnitude of analytical error that may be present in the final patient result. It is useful to express the errors in the same units as the patient result so that the effect can be readily observed. The experiments may be performed in whatever order provides the critical information in the quickest manner. If the results from a single experiment demonstrate that the errors are too large to be acceptable, the method must be modified to reduce the errors or it must be rejected. For a method to be acceptable, all experiments must demonstrate that the analytical errors are acceptably small. Specific recommendations for the performance of experimental studies will be given in a forthcoming document from the Scientific Committee of IFCC.

2.4 Performance standards

Acceptable performance requires that the analytical errors be small, below some defined limit of allowable error. The definition of that limit of allowable error requires the professional judgment of the evaluator (or evaluating group) and he must carefully consider the clinical context in which the method will be used, (e.g. emergency clinical work, population screening, research) as well as sources of information for "state of the art" performance, "medical usefulness" requirements, and requirements based on the variation of the population as well as the variation of an individual. The use of calculated analytical goals must be moderated as analytical goals must be medically realistic and attainable technically (5).

2.5 Criteria

An objective judgment on performance will require that criteria be defined such that the observed errors can be compared with the defined allowable limit of error.

2.5.1. Analytical confidence limits. It is desirable to make the decision on performance with a known level of confidence, in order to be sure that the data are sufficient to support a decision of either acceptance or rejection. This can be done by calculating a confidence interval for the observed estimate of analytical error.

2.5.2. Statistical estimates of error. This can be done by comparing the statistical estimate of an error with the limit of allowable error. If the observed error is smaller than the specified limit of allowable error, then performance is acceptable. If the observed error is larger than the specified limit of allowable error, then the performance is not acceptable.

2.5.3. Data collection. It must be recognized that a method whose performance is marginal with respect to the stated limit of allowable error may require excessive numbers of tests to demonstrate whether it is acceptable or not. It is desirable to set an upper bound for the number of tests to be performed, and if a method has not been demonstrated to be unacceptable when these tests are completed, then it is considered to be acceptable.

3. Application of guidelines to clinical chemistry kits

In the area of clinical chemistry, testing protocols of this nature are becoming increasingly well defined. These vary in sophistication and rigor, and in practicability. The attempt in this document is to define the outline of a practical protocol that could be implemented in a variety of laboratories. Specific recommendations for experimental work will be the subject of a future document.

3.1 Practicability characteristics

These are of primary importance in selecting methods which meet the needs of the laboratory and which can be implemented in the laboratory of interest.

The following lists the practicability characteristics which must be considered:

3.1.1. Specimen.

Type
Volume
Type of anticoagulant
Stability
Nature of Container

3.1.2. Operating requirements.

Instructions
Instrument
Additional equipment
Reagent information (formulation, stability, etc.)

3.1.3. Service requirements.

Power supply
Ventilation
Environment
Reagent disposal

3.1.4. Consumables requirement.

Reagents
Sample containers
Disposable/replaceable instrument components

3.1.5. Staff.

Technical skill
Training

3.1.6. Cost

Cost per test

3.1.7. Speed of analysis

Analyst time per test
Rate for single sample
Rates for varied numbers of multiple samples

3.1.8. General

Safety aspects
Equipment maintenance
Availability of product
Manufacturer's support

3.2 *Reliability characteristics*

These generally include analytical range, precision, recovery, interference, specificity, accuracy, and detection limit. These terms have been defined by IFCC (3). The words in the brackets have been added by this committee for additional clarification.

3.2.1. Analytical range. The range of concentration or other (measured) quantity in the specimen over which the method is applicable without modification. The cut off concentration(s) of a coarse-graduated scale test should be well defined and correspond to specified clinical needs, cf. *3.2.9.*

3.2.2. Precision. Agreement between replicate measurements. It has no numerical value. See imprecision.

3.2.3. Imprecision. Standard deviation or coefficient of variation of the results in a set of replicate measurements. The mean value and number of replicates must be stated and the design used must be described in such a way that other workers can repeat it. This is particularly important whenever a specific term is used to denote a particular type of imprecision, such as between-laboratory, within-day, or between-day.

3.2.4. Recovery. Refers to the ability of the method to correctly measure the amount of pure analyte added to the original sample.

3.2.5. Interference. The effect of a component (other than the one to be measured) on the accuracy of measurement of another component.

3.2.6. Specificity. The ability of an analytical method to determine solely the component(s) it purports to measure. It has no numerical value. It is assessed on the evidence available on the components which contribute to the result, and on the extent to which they do.

3.2.7. Accuracy. Agreement between the best estimate of a quantity and its true value. It has no numerical value. See inaccuracy.

3.2.8. Inaccuracy. Numerical difference between the mean of a set of measurements and the true value. This difference (positive or negative) may be expressed in the units in which the quantity is measured, or as a percentage of the true value. (The true value may be defined for operational purposes as that value determined by a specified definitive or reference method.) The inaccuracy (and also, the imprecision) should be given at several clinically important levels.

3.2.9. Detection limit. The smallest single result which, with a stated probability (commonly 95%), can be distinguished from a suitable blank. The limit may be a concentration or an amount and defines the point at which the analysis becomes just feasible. Binary tests should be designed in such a way that the clinically relevant discrimination limit concentration is close to the detection limit. Because of some imprecision of the test, a certain number of false positive test results and false negative test results is inherent. The test should be designed in such a way that the discrimination limit concentration minimizes the clinical consequences of false positive and false negative test results.

4. Summary

The purpose of the document is to state the policy of IFCC on the evaluation of clinical chemistry kit methods. Subsequent documents will provide specific guidance on experimental procedure, the interpretation of data and criteria for acceptable performance.

5. References

1. Rubin, M., Barnett, R. N., Bayse, D., Beutler, E., Brown, S. S., Logan, J. E., Reimer, C. B., Westgard, J. O. & Wilding, P. Provisional recommendation (1978) on evaluation of diagnostic kits.
Part 2. Guidelines for the evaluation of clinical chemistry kits. Clin. Chim. Acta (1979) *95*, F163–F168, J. Clin. Chem. Clin. Biochem. (1979) *17*, 622–664, Clin. Chem. (1979) *25*, 1503–1505.
2. Rubin, M., Barnett, R. N., Bayse, D., Beutler, E., Brown, S. S., Logan, J. E., Reimer, C. B., Westgard, J. O. & Wilding, P. Revised recommendation (1983) on evaluation of diagnostic kits.
Part 1. Recommendations for specifications on labelling of clinical laboratory materials. J. Clin. Chem. Clin. Biochem. *21*, 893–898.
3. Büttner, J., Borth, R., Boutwell, J. H., Broughton, P. M. G. & Bowyer, R. C. Approved recommendation (1978) on quality control in clinical chemistry.
Part 2. Assessment of analytical methods for routine use. Clin. Chim. Acta (1979) *98*, 145F–162F, J. Clin. Chem. Clin. Biochem. (1980) *18*, 78–88.
4. Guiding principles and recommendations on labelling of clinical laboratory materials (1978). WHO Memorandum. Bull. WHO (1978) *56*, 881–885.
5. Westgard, J. O. (1977) The development of performance standards and criteria for testing the precision and accuracy of laboratory methods. In Proceedings of the 1976 Aspen Conference on Analytical Goals in Clinical Chemistry, American College of Pathologists, Chicago, IL USA

Summary of IFCC/WHO Documents Produced by the Expert Panel on Evaluation of Diagnostic Reagent Sets

J. Logan

The following IFCC/WHO documents were published early in 1983 by WHO:

Lab/83.1
IFCC/WHO Principles and Recommendations on Evaluation of Diagnostic Reagent Sets used in Health Laboratories with Limited Resources
Part 1. Selection and Evaluation using Reference Materials – General Considerations
Part 2. Evaluation of Performance using Reference Materials of Analytes Commonly Determined in Blood Serum or Plasma

Lab/83.2
Evaluation of Performance of Diagnostic Reagent Sets Used in Health Laboratories with Limited Resources
Appendix A: Glucose

Lab/83.3
ibid
Appendix B: Urea or Carbamide

Lab/83.4
ibid
Appendix C: Total Bilirubin

Lab/83.5
ibid
Appendix D: Total Protein

Lab/83.6
ibid
Appendix E: Albumin

This group of documents was prepared by the Expert Panel on the Evaluation of Diagnostic Reagent Sets and the work to develop them received financial support from the World Health Organization. The documents were designed to be used primarily in developing countries in health laboratories with limited resources. It is conceivable that smaller laboratories within these countries might undertake the preliminary evaluations described therein and thus eliminate the poorer performing products. On the other hand, the complete assessment would most likely have to be carried out on behalf of others in a central laboratory that possesses the necessary expertise and facilities.

The Part 1 document describes how kits may be initially selected for evaluation by first reviewing and assessing labelling and product information provided by the manufacturer and secondly by checking how closely the features of the kit match the specific needs and available resources of the user laboratory. This is accomplished by means of a rating sheet for the former and a checklist of the practicability characteristics included in the latter. The general considerations to evaluate the analytical performance of a method are outlined with experiments designed to estimate the magnitude of analytical error that may be present in the final patient result. The performance requirements for acceptability are discussed so that when results from preliminary experiments produce errors that are too large, the kit can be rejected without further assessment. The strategy of the evaluation is outlined and consists of three parts:

1) a preliminary examination of precision, linearity and accuracy, 2) a measurement of bias effects of specific interferants, and 3) a more uniform extended study to provide estimates of linearity, within-run and among-run imprecision, and bias versus reference material values. The various evaluation steps are performed sequentially and are detailed in the Part 2 document together with the appropriate analyte appendix.

The Part 2 document proper describes in detail those steps in the evaluation scheme common to the five analytes listed in the appendices A to E inclusive. Data sheets (Form 1, 2, 3-1, 3-2, 3-3, 4-1, 4-2, 4-3, 4-4, 5 and 6) are provided for recording results of each evaluation step and applying test criteria to determine the acceptability of performance. The preliminary evaluation requires making measurements on both a series of calibration solutions of the analyte and on reference sera. This requires about 100 tests, and by using simple criteria for precision and accuracy it may be determined whether to reject the product or continue the assessment. The linearity of the test and the stability of the kit reagents can also be determined from the preliminary data. Procedures for 40 to 60 kit analyses are required to test the amount of error caused by known or suspected interfering substances or specimen characteristics. The final extended study consists of analytical runs on some 30 samples on six separate days. Both calibrator solutions and reference sera are used which measure bias and imprecision of the kit procedure. About 200 kit tests are required for this. The test criteria to be applied to all results are described as well as the statistical treatment to be used to analyze the imprecision. A final decision concerning acceptance or rejection of the kit is made after inserting all test results in the summary of the evaluation form (Form 6).

The individual analyte appendices, A: glucose, B: urea or carbamide, C: total bilirubin, D: total protein and E: albumin, describe the specific instructions required to carry out the evaluation when a kit designed for measuring the particulate analyte is being evaluated. The scientific basis of the analysis is discussed, the materials needed in the laboratory (e.g., volumetric glassware, type of instrument) are specified, directions for preparation of calibration solutions and reference sera of varying concentrations and finally full details on all the steps to carry out the preliminary trial, the interference and extended evaluation studies. Test criteria are applied which relate to the acceptable error for the analyte. Exemplary data are provided on the data forms for each analyte so that the reader may follow the process in finally determining the acceptability or rejection of the product.

Prior to publication of the IFCC/WHO documents, a field trial of the Parts 1 and 2 documents and the appendix for glucose was conducted in 1981 in five developing countries. This was coordinated by WHO with a commercial glucose kit and reference materials being provided. The study indicated the suitability of the documents for use in developing countries. These are now available from the Health Technology Unit, WHO, Geneva, Switzerland or from any WHO Regional Centre. They may also be ordered from the IFCC Technical Secretariat, P.O. Box 986, DK-2400 Copenhagen NV, Denmark.

VII. Expert Panel* on pH and Blood Gases

* A. H. J. Maas, Utrecht (NLD), Chairman
R. A. Durst, Washington, DC (USA) until 1982–05
O. Siggaard-Andersen, Copenhagen (DK)
H. F. Weisberg, Milwaukee (USA), Secretary from 1982–05
W. G. Zijlstra, Groningen (NLD)

Reference Method for pH Measurement in Blood[1])

Prepared for publication by A. H. J. Maas, H. F. Weisberg, W. G. Zijlstra, R. A. Durst and *O. Siggaard-Andersen*

Comments from the viewpoint of language other than English are encouraged.

Contents

1. Introduction

The purpose of describing a reference method for pH measurement in blood is to provide a method to establish reliably the pH of blood samples on the basis of the primary reference materials. These are aqueous buffer solution, the pH values of which have been established by means of the definitive method for pH measurement in dilute aqueous solutions (see § 2.3.).

2. Principles and Definitions

2.1. pH is defined as the negative decadic logarithm of the relative molal activity of hydrogen ions:

$$pH = -\lg a_{m,H^+} = -\lg(\underline{\gamma}_{m,H^+} \cdot m_{H^+}/m^{\ominus}] \qquad \text{Eq. 1}$$

where a_{m,H^+} is the relative molal activity, $\underline{\gamma}_{m,H^+}$ is the molal activity coefficient and m_{H^+} is the molality of the hydrogen ions and $m^{\ominus} = 1$ mol/kg.

2.2. pH measurement in a solution may be performed by several methods based on different analytical principles: colorimetry (colour indicator), calculation (e. g. from pCO_2 and $cHCO_3^-$ by the *Henderson-Hasselbach* equation), and potentiometry with different types of electrodes (hydrogen gas, quinhydrone, metal oxide, glass and ion-sensitive field effect transistor). For pH measurement in blood (2), the potentiometric method using a glass electrode is the method of choice.

2.3. The definitive method (3, 4) for pH measurement in dilute inorganic aqueous solutions is based on measuring the potential of a cell consisting of a hydrogen gas electrode and a silver/silver chloride reference electrode without a liquid/liquid junction (cell without transference). This cell is often

[1]) Stage 2, Draft 2; 1982-12 with a view to a Recommendation

called the *Harned* cell. The definitive method is employed by reference laboratories, e. g., the National Bureau of Standards (NBS, USA) which has assigned pH(S) values to a series of primary calibration solutions in the pH range from 3 to 11 with an estimated uncertainty in pH of ± 0,005 (see also § 11).

2.4. The IUPAC reference method (5) for pH measurement in aqueous solutions is based on the use of a cell consisting of a hydrogen gas electrode and a reference electrode R_1 with a concentrated KCl liquid/liquid junction according to the scheme:

$$R_1 \mid KCl\,(m \geq 3{,}5\ mol/kg) \,\vdots\vdots\, solution\ X \mid H_2\,(g) \mid Pt \qquad (cell\ a)$$

The vertical lines represent phase boundaries; the double dotted vertical line indicates the liquid/liquid junction between the two different solutions.

A small unknown inaccuracy is introduced by the uncertainty of the residual liquid/liquid junction potential.

The hydrogen gas electrode is not suitable for pH measurements in blood due to protein contamination of the platinum surface and potential interferences by redox couples, e. g. oxyhaemoglobin/reduced haemoglobin (6).

2.5. The proposed IFCC reference method (1) for pH measurement in blood and other biological fluids is based on the use of a cell consisting of a pH glass electrode and a reference electrode R_1 with a saturated potassium chloride liquid/liquid junction according to the scheme:

"Reference Electrode" "Glass Electrode"

$$R_1 \mid KCl\,(satd.) \,\vdots\vdots\, sol.\ X \mid G \mid inner\ ref.\ sol. \mid R_2 \qquad (cell\ b)$$

The glass electrode is presented as a composite cell. On each side of the glass wall (G) an electrical surface potential develops across the glass/solution boundary, which varies linearly with pH over a wide range. The inner side of the glass electrode is in contact with a reference solution of constant pH. The potential difference across the membrane is a function of the pH of the sample solution X (calibration solution or blood) (2, 3). Electrodes R_1 and R_2 are connected to the pH-meter. The cell is calibrated by means of the primary calibration solutions of the NBS pH(S) scale mentioned in § 2.3., and described in detail in § 11.

2.6. Routine methods for pH measurement in blood are evaluated by the reference method described in § 2.5. Most routine methods are based on cell b and are usually calibrated with secondary calibration solutions. Secondary calibration standards are aqueous buffer solutions, the pH of which have been established by means of the reference method (see § 2.4.) and may be used for calibration of routine methods. They often produce highly precise, but not necessarily accurate pH data because of variations in electrode systems, liquid/liquid junctions, calibration and measurement procedures. In order to monitor the accuracy of routine pH measurements it is necessary to employ quality control solutions, e. g. blood samples for which the pH value has been established with the reference method (see § 5.2. and § 9.1.).

3. Instrumentation and Equipment

The reference system comprises a pH glass electrode and a reference electrode with a saturated potassium chloride liquid/liquid junction; the whole system is at 37 °C (see fig. 1).

3.1. The glass electrode

3.1.1. The *sensitivity* of the glass electrode is defined as:

$$g \stackrel{def}{=} \frac{dE}{d\lg a_{m,H^+}} = -\frac{dE}{dpH} \qquad \text{Eq. 2}$$

where E is the cell potential.

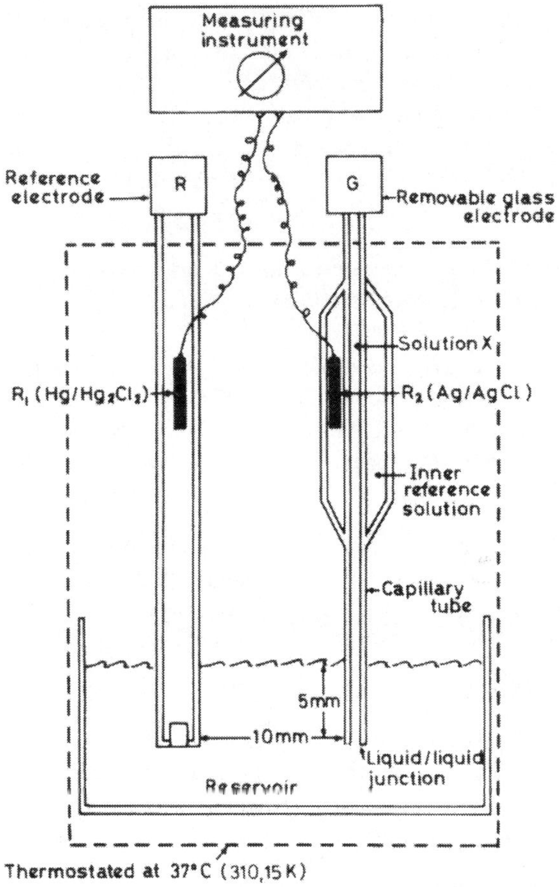

Figure 1. Schematic presentation of the pH measuring system.

The theoretical *Nernst*ian sensitivity of an ideal H^+ electrode is:

$$g^\circ = \frac{R \times T \times \ln 10}{F} = 0{,}0615\ J \times C^{-1}$$
$$= 0{,}0615\ V \qquad \text{Eq. 3}$$

where R is the gas constant ($8{,}3134\ J \times K^{-1} \times mol^{-1}$), T is the absolute temperature ($T = 310{,}15$ K)[2]) and F is the *Faraday* constant ($96475\ C \times mol^{-1}$).

[2]) 310,15 K = 37 °C.

The *relativity* of the glass electrode is defined as:

$$S = g/g^{\circ} = -\frac{dE}{dpH} \times \frac{F}{R \times T \times \ln 10}$$

$$= \frac{dE}{dpH} \times 16{,}25 \qquad \text{Eq. 4}$$

The relative sensitivity (S) of the glass electrode should not deviate by more than $\pm$ 0.02 from the theoretical value (1,00). This may be measured as described by *Camoes* (7), *Covington* (8) and *Sprokholt* et al. (9).

3.1.2. The *selectivity* of the glass electrode for H^+ should be established. Na^+ is the only cation which may cause a significant interference (3). This so-called *alkaline error* causes a lower pH than the true value in the presence of Na^+ and this bias increases with increasing pH, sodium ion activity and temperature. For the reference method, the sodium error of the pH value should be less than 0,003 (or $\Delta E < 0{,}2$ mV) at pH = 7,6 with a sodium ion concentration of 0,15 mol/l at 37 °C.

A method for measuring the sodium error is described by *Covington* (8) and *Sprokholt* et al. (9).

3.1.3. The *response time* of the electrode chain with alternating measurements on both the calibration solutions should be less than 15 s for 95% response when the sample is preheated to 37 °C to avoid thermal effects.

3.1.4. The *stability* should be such that the drift of the electrode chain does not exceed 0,06 mV $\times$ h^{-1} (pH drift < 0,001 $\times$ h^{-1}) when the electrode remains filled with the calibration solution: 1 : 3,5 phosphate buffer (see § 4.1.).

3.1.5. The *resistance* of the glass electrode should be less than 10^{-3} times electrometer-input-impedance, in order to achieve measuring results accurate within 0,1%.

3.1.6. *Geometry:* the glass electrode has been constructed with various designs for special applications. The capillary type is chosen for the reference method of pH measurement in blood. The inner diameter should be between 0,3 and 0,6 mm to ensure a reasonable ratio between blood volume and surface area of the glass membrane and to allow temperature measurement within the capillary.

The length should be between 4 and 10 cm for practical reasons.

Other geometrical designs, e. g. a bulb shape, may be employed for the reference method, provided the volume of the cuvette, wash out, possibility of "anaerobic" filling and temperature control are equivalent to the capillary type electrode.

3.1.7. The *inner reference solution* should be a solution having a stable pH and a stable chloride activity into which the internal reference silver/silver chloride electrode is completely submerged.

3.1.8. *A memory effect* of the glass electrode is observed as an erroneously low or high pH value when changing the sample from a weakly buffered solution with a low pH to another with a high pH or vice versa.

This is due to a certain buffer capacity of the glass membrane and may vary with the type of glass. This may be overcome by flushing the electrode capillary 5 times with the new sample with each sample to remain in the capillary for 15 to 30 s. No air should be introduced between the flushings.

3.1.9. Protein coatings on the glass membrane change the membrane characteristics and slow down the electrode response. For measurements on blood, cleaning of the glass electrode with a protein removing agent is recommended (see § 4.4.).

3.2. The reference electrode

3.2.1. The *mercury/calomel electrode* is the reference half-cell which should be used as reference electrode R_1 (see cell a in § 2.4. and cell b in § 2.5.). The mercury/calomel electrode provides its own liquid junction by use of a saturated potassium chloride bridge solution.

3.2.2. The *silver/silver chloride* electrode is commonly used as reference electrode R_2 (see cell b in § 2.5.).

3.3. The liquid/liquid junction

3.3.1. The *bridge solution* is a potassium chloride solution saturated at the measuring temperature of 37 °C.

The *liquid/liquid junction potential* between the bridge solution and the sample solution X (calibration solution or blood) should be formed just before measuring the cell potential. The saturated potassium chloride solution does not completely eliminate *the residual liquid junction potential*, i. e., the difference between the junction potential with the calibration solution and the junction potential with blood (2). The resulting bias is small (< 2 mV) and cannot be measured or eliminated on theoretical grounds.

3.3.2. The *temperature dependence* (10, 11) of the residual liquid/liquid junction potential,

"saturated KCl sol. ⦙⦙ calibration sol. (or blood)",

is 0,05 mV $\times$ K^{-1} (in terms of pH: 0,001 $\times$ K^{-1}), due to differences in activity and mobility of K^+ and Cl^- at different temperatures. To avoid temperature effects upon the residual liquid/liquid junction potential both electrodes of the pH cell including the liquid/liquid junction should be maintained at the same temperature (2).

3.3.3. The *effect of erythrocytes* on the junction potential ions at the boundary,

"saturated potassium chloride solution ⦀ blood",

results in a lower pH value than that of the corresponding plasma when the pH of blood is measured with this reference system. The pH bias depends on the volume fraction of erythrocytes in the blood (10, 11), with a hematocrit of 0,45 the bias is 0,01 pH. The true plasma pH value may be obtained by avoiding the presence of erythrocytes at the junction (see § 6.3.3.).

3.3.4. The *geometry* of the liquid/liquid junction also influences the junction potential. A stable and reproducible liquid junction is obtained with the free diffusion type of junction when it is arranged in such a way that the higher density potassium chloride solution is below the test solution; a capillary tube less than 0,4 mm diameter, dipping at least 5 mm into the saturated potassium chloride bridge with a distance of at least 10 mm to the calomel reference electrode R_1 to avoid contamination, is recommended (see fig. 1).

3.4. The pH-meter

The pH-meter is a high impedance voltmeter which can measure the electromotive force of the electrochemical pH-cell.

A suitable instrument should have the following properties and specifications:

1. Input bias current (I_b) should be less than 5×10^{-13} A (25 °C).
2. Input bias current change with temperature ($\Delta I_b/\Delta T$) should be less than 5×10^{-14} A $\times$ K^{-1} at ambient temperature below 30 °C.
3. Input resistance should be at least 10^{12} Ω at the glass electrode input.
4. Input capacitance should be less than 10^{-9} F.
5. The pH-meter should be floating with respect to ground; resistance to ground should be more than 10^8 Ω.
6. Zero point stability should be better than 50 μV $\times$ K^{-1} and slope stability better than 10^{-4} μV $\times$ K^{-1}.
7. There should be a mode switch to select reading of E(mV) or pH.
8. Resolution should be at least 0,1 mV in E-mode or 0,001 in pH-mode.
9. Slope error in E-mode should be less than 0,05% at 25 °C ambient temperature.
10. An isopotential control for adjusting the zero point of the pH-meter to the pH value of the first calibration solution (see § 6.2.).
11. A calibration control for adjustment of the pH reading to the pH value.
12. A sensitivity (slope) control for adjustment of the reading to the pH value of the second calibration solution (see § 6.2.).
13. The instrument should preferably have a digital display and outputs for recorder and for printer.

3.5. Temperature control

The whole cell should be maintained at a temperature of 37,0 °C (± 0,05 °C) and it should be possible to measure the temperature inside the cuvette (i. e. the pH sensitive glass capillary) by means of a thermistor. It is recommended to calibrate the thermistor against a certified thermometer.

4. Reagents

4.1. Calibration solutions

Primary reference materials are issued as salts by the NBS and the user is required to prepare the calibration solutions according to a prescribed procedure. For pH measurement of blood, two reference solutions have been certified, with a 1 : 1 molality ratio of potassium dihydrogen phosphate: disodium hydrogen phosphate (pH = 6,839 at 37 °C) and a 1 : 3,5 molality ratio (pH = 7,392 at 37 °C). For details see Appendix § 11.)

4.2. Saturated potassium chloride solution

Saturated potassium chloride solution for the reference electrode and liquid/liquid junction should be prepared from analytical grade potassium chloride and deionized water at 37 °C.

4.3. Sodium chloride solution

A sodium chloride solution (0,16 mol/l) for flushing the electrode system before and after introducing the blood sample should be prepared from analytical grade sodium chloride and deionized water.

4.4. Cleaning solutions

For cleaning of the electrode assembly and removal of protein contamination, a solution containing an appropriate proteolytic enzyme in a neutral solvent or commercially available detergents or hypochlorite solution (1 mol/l) is recommended.

5. Specimen Procurement and Processing for Quality Control Purposes

5.1. Procurement of whole blood for pH measurement

Whole blood is collected in a glass syringe (5 ml) containing a flat stainless steel plate (diameter 0,2 mm less than diameter of plunger, thickness 0,5 mm) for mixing. The dead space of the syringe which should be less than 3% of the total volume, is filled with a sterile, neutral, isotonic heparinate solution (10^6 IU/l ≅ 10 g/l) and the inside of the barrel is wetted by moving plunger up and down (2). The concentration of heparin in the blood sample should be 20.000–100.000 I.U./l.

The blood pH may be altered, e.g. by tonometry or adding some acid or base to produce different control samples with abnormal pH. See § 9.1. for the preparation of blood as a matrix quality control sample.

5.2. Performance of control measurements

The pH measurements by means of the reference method and the method to be controlled are repeated alternatingly at least twice with as little time interval as possible.

6. Measurement Procedure

A typical protocol is described, details of which may vary according to the electrode system and pH-meter used. Detailed instructions should be found in the manuals of the manufacturer.

6.1. Instrumental

1. The pH-meter is turned on to reach thermal equilibrium.
2. The electrode system is checked for cleanliness and proper temperature (37 °C).

6.2. Calibration

1. The zero point of the pH-meter is adjusted precisely to the third decimal with the isopotential control to the assigned pH(S) value of the calibrator 1 (1 : 3,5 phosphate buffer).
2. Before filling the glass electrode is effectively rinsed by alternately aspirating air and calibrator 1; the pH value is adjusted with the *calibration adjustment control.* The adjustment is checked twice by aspirating more calibrator 1 without air in between; imprecision for pH 0,001 (≈ 0,06 mV).
3. The glass electrode is then rinsed and filled in a similar fashion with the calibrator 2 (1 : 1 phosphate buffer). The pH value is adjusted accurately with *the gain adjustment control.* The adjustment is also checked twice without air in between; imprecision for pH 0,001 (≈ 0,06 mV).
4. Recheck calibrator 1 according to step 2.
5. The gain calibration should be checked periodically.

6.3. Measurement

1. The glass electrode is flushed by alternately drawing air and sodium chloride solution (see § 4.3.) at least five times in order to rinse the buffer off the glass surface.
2. The electrode is slowly filled with well mixed blood in excess to overcome the memory effects. Avoid any air in the electrode during the filling procedure.
3. Next, a few µl of plasma are aspirated into the capillary tip to be interposed between the blood and the saturated KCl to avoid the effect of the blood cells on the liquid/liquid junction potential (see § 3.3.3.). This plasma should be obtained from the original blood specimen. It is important that no air bubble is interposed between the blood and plasma.
4. After 20 s the two half cells are connected by establishing the liquid/liquid junction; wait 2–3 s and read pH.
5. The blood is flushed out with saline and then the glass electrode is flushed by alternately drawing air and sodium chloride solution at least five times, in order to rinse the blood off the glass surface.
6. Recheck calibrator 1 according to § 6.2.2. The pH difference should be less than 0,005, otherwise the whole measurement procedure should be repeated.

7. Analytical Variability

7.1. Precision

The following results (SD) exemplify the imprecision which should be obtained with the reference method:

Within day (n = 20)		*between days* (n = 20)	
aqueous control	0,002	aqueous control	0,003
plasma	0,003		
blood	0,005		

7.2. Accuracy

Inaccuracy (Δ) is defined as the difference between the mean value of a series of measurements and the true value. The true pH value is $-\lg a_{m,H^+} = -\lg (\gamma_{m,H^+} \cdot m_{H^+}/m^{\ominus})$ with the best theoretical estimation for γ_{m,H^+}. Based on a comparison of the various possibilities for calculating γ_{m,H^+}, *Bates* (3) estimated that the definitive method for pH measurement in primary calibration solutions using the *Bates-Guggenheim* convention provides the true $-\lg a_{m,H^+}$ with a maximum inaccuracy of ± 0,010.

Further, the inaccuracy of the pH value of a primary calibration solution prepared according to the described method may amount to 0,005. Thus the overall inaccuracy for the pH of a primary calibration solution is maximally ± 0,015.

Due to the residual liquid/liquid junction potential, however, a pH uncertainty of ± 0,010 remains, which has to be added to the above mentioned inaccuracy. Therefore an inaccuracy, estimated to be at most ± 0,025 with regard to $-\lg a_{m,H^+}$, may be expected for a blood pH measured by the reference method.

7.3. Sources of error

The most common errors are the following:

a) inaccurate calibration solutions (improper preparation, carbon dioxide absorption, microbiological contamination)
b) inadequate temperature control, i. e., a deviation of more than ± 0,1 °C from 37 °C
c) contamination of the electrode system with protein
d) inadequate flushing or filling of the electrode system with calibration solution or blood
e) the effect of the erythrocytes on the residual liquid/liquid junction potential (10, 11)
f) faulty electrical grounding in the analytical system
g) defective electrodes

8. Secondary Calibration Solutions (12)

8.1. Secondary calibration solutions

Secondary calibration solutions are produced by commercial organizations or by individual laboratories for use as "working standards".

The pH accuracy of these buffers may be in error by several hundredths, although the assigned inaccuracy is usually given as $|\Delta \text{pH}| < 0{,}005$. Ideally these secondary reference materials are directly related to the primary calibration solutions by the present reference method.

8.2. The 1 : 4 *Sörensen* phosphate buffer is often used as a secondary calibration standard (13). "The pH value measured with the definitive method and thus consistent with the NBS pH scale is 7,381 at 37 °C" (10). Because of higher salt concentration (ionic strength I = 0,18 mol/kg), buffer capacity and stability, in practice this calibration solution is superior to the primary 1 : 3,5 phosphate buffer, which is a primary NBS calibration standard.

8.3. A phosphate-chloride buffer composed of potassium dihydrogen phosphate 0,005217 mol/l, disodium hydrogen phosphate 0,018258 mol/l, and sodium chloride 0,1 mol/l with ionic strength I = 0,16 mol/kg has recently been described (14).

The purpose of this buffer is to obtain a minimal liquid/liquid junction potential. A disadvantage is the lower phosphate concentration and hence the buffer value is lower than the certified NBS 1 : 3,5 phosphate buffer.

9. Quality Control Solutions (12)

9.1. For measuring blood pH, whole blood with a known pH is the ideal solution for quality control. Blood employed for assessing the accuracy of a routine method, should be devoid of leukocytes to reduce the rate of oxygen consumption and carbon dioxide production of the sample.

This is accomplished by centrifuging the blood (1000 g ≅ 10000 N/kg for 10 min), removing the leukocyte layer by aspiration and resuspending the erythrocytes in the plasma. The blood may then be equilibrated in a tonometer with any desired gas mixture before pH measurement by the reference method. However there are many problems involved in making whole blood a stable quality control material especially for the gas quantities.

9.2. For these reasons simulated quality control materials with assigned pH, pCO_2 and pO_2 values have been developed which are suitable for quality control of modern instruments by allowing simultaneous measurement of all three quantities. At the present time, these blood pH/gas quality control materials include gas equilibrated solutions, the pH of which should be established with this reference method.

Such solutions are prepared from:

a) whole blood (15)
b) stabilized blood cell suspensions (16)
c) frozen haemolysed red cell solution (17)
d) aqueous preparation of lyophilized serum (18)
e) fluorocarbon emulsions with organic/inorganic buffer and $NaHCO_3$ (I = 0,16 mol/kg) (19)
f) various organic buffer solutions with $NaHCO_3$ (I = 0,16 mol/kg) (20)
g) aqueous solutions of $NaHCO_3$, NaCl, KH_2PO_4 and Na_2HPO_4 (I = 0,16 mol/kg) (21, 22)
h) aqueous solutions of $NaHCO_3$ and NaCl (I = 0,16 mol/kg) (23)

Each of these materials has its own particular advantages and limitations, and it is expected that further development in this area will occur.

10. References

1. Siggaard-Andersen, O., Durst, R. A. & Maas, A. H. J. (1980) Physico-chemical quantities and units in clinical chemistry. Clin. Chim. Acta *108*, 501F–539F; J. Clin. Chem. Clin. Biochem. *18*, 829–854.
2. Siggaard-Andersen (ed.) (1974) The Acid-Base Status of the Blood, 4th revised edition, Chapter 4, pp. 145–165, Munksgaard, Copenhagen.
3. Bates, R. G. (ed.) (1973) Determination of pH. Theory and Practice, 3rd ed., Chapter 4, pp. 59–104, John Wiley and Sons, New York.
4. Durst, R. A. (ed.) (1975) Standard Reference Materials: Standardization of pH measurements, NBS Special Publication 260–53, pp. 1–38, US Government Printing Office, Washington D.C.
5. Whiffen, D. H. (ed.) (1979) Manual of Symbols and Terminology for Physico-chemical Quantities and Units. IUPAC, Division of Physical Chemistry, Commission on Symbols, Terminology and Units. 1978 edition. Pure & Appl. Chem. *51*, 30–31.
6. Warburg, E. J. (1922) Carbonic acid compounds and hydrogen ion activities in blood. A contribution to the theory of the equation of L. J. Henderson and K. A. Hasselbalch. Biochem. J. *16*, 153–340.
7. Camoes, M. F. G. F. C. & Covington, A. K. (1974) New procedure for calibrating glass electrodes. Anal. Chem. *46*, 1547–1551.
8. Covington, A. K. & Ferra, M. I. A. (1977) Buffer solutions for testing glass electrode performance in aqueous solutions over the pH range 0–14 at 25 °C. Anal. Chem. *49*, 1363–1368.
9. Sprokholt, R., Maas, A. H. J., Rebelo, M. J. & Covington, A. K. (1982) Determination of the performance of a glass electrode in aqueous solution at 37 °C, in the physiological pH-range (6.4–7.6) and at the physiological sodium-ion concentration (0.15 mol/l). Anal. Chem. *54*, 53–59.
10. Maas, A. H. J. (1970) pH determination of body fluids with a micro glass electrode and a saturated KCl bridge in the cell. Clin. Chim. Acta *28*, 373–390.
11. Salling, N. & Siggaard-Andersen, O. (1971) Liquid/liquid junction potentials between plasma or erythrolysate and KCl solutions. Scand. J. Clin. Lab. Invest. *28*, 33–40.
12. Durst, R. A., Siggaard-Andersen, O. & Maas, A. H. J. (1981) Reference and quality control materials for pH and blood gases. In: Proceedings of the 5th meeting of the IFCC Expert Panel on pH and blood gases held at Copenhagen, June 16–18, 1980 (Siggaard-Andersen, O., ed.) pp. 89–100.
13. Spinner, M. B. & Kokholm, G. (1961) Determining of the pH of a phosphate buffer solution for blood measurements. Scand. J. Clin. Lab. Invest. *13*, 1–7.
14. Bates, R. G., Vega, C. A. & White, D. R. (1978) Standards for pH measurements in isotonic saline media of ionic strength $I = 0.16$. Anal. Chem. *50*, 1295–1300.
15. Leary, E. T., Delaney, C. J. & Kenny, M. A. (1977) Use of equilibrated blood for internal blood-gas quality control. Clin. Chem. *23*, 493–503.
16. Quantra Whole-Blood Gas Control, DADE Division American Hospital Supply Corp., Miami, FL 33152.
17. Steiner, M. S., Shapiro, B. A., Kavanaugh, J., Walton, J. R. & Johnson, W. (1978) A stable Blood Product for pH-Blood gas Quality Control. Clin. Chem. *24*, 793–795.
18. Versatol Acid Base, General Diagnostics, Division of Warner Lambert Co., Morris Plains, NJ 07950.
19. Feil, M. C., Legg, K. D. & Cormier, A. D. (1981) A blood gas control with improved oxygen buffering capacity. Clin. Chem. *27*, 1111.
20. Komjathy, Z. L., Mathies, J. C., Parker, J. A. & Schreiber, H. A. (1976) Stability and precision of a new ampuled quality-control system for pH and blood-gas measurements. Clin. Chem. *22*, 1399–1401.
21. Veefkind, A. H., van den Camp, R. A. M. & Maas, A. H. J (1975) Use of carbon dioxide and oxygen-tonometered phosphate – bicarbonate – chloride – glycerol – water mixtures for calibration and control of pH, pCO_2 and pO_2 electrode systems. Clin. Chem. *21*, 685–693.
22. Maas, A. H. J., Veefkind, A. H., van den Camp, R. A. M., Teunissen, A. J., Winckers, E. K. A. & Jansen, A. P. (1977) Evaluation of Ampouled Tonometered Buffer solutions as a Quality Control System for pH, pCO_2 and pO_2 Measurement. Clin. Chem. *23*, 1718–1725.
23. Noonan, D. C. & Burnett, R. W. (1974) Quality-Control System for Blood pH and Gas measurements with use of a Tonometered Bicarbonate-Chloride Solution and Duplicate Samples of Whole Blood. Clin. Chem. *20*, 660–665.

11. Appendix

U. S. Department of Commerce
Maurice H. Stans
Secretary

National Bureau of Standards
L. M. Branscomb, Director

Certificate

Standard Reference Materials

Potassium Dihydrogen Phosphate (186-I-c)

Disodium Hydrogen Phosphate (186-II-c)

These standard reference materials are intended for use in defining the pH(S) scale. The lots of potassium dihydrogen phosphate (KH_2PO_4) and disodium hydrogen phosphate (Na_2HPO_4) were prepared to ensure high purity and uniformity. They meet the specifications of the American Chemical Society for reagent-grade materials, but should not be considered as entirely free from impurities such as traces of water, free acid or alkali, carbon dioxide, chlorides, sulfur compounds and heavy metals.

The pH(S) values listed below correspond to $\log(1/a_H)$, where a_H is a *conventional* activity of the hydrogen (hydronium) ion referred to the standard state on the scale of molality. The values were derived from the emf of cells without liquid junction by the method of calculation described in the Journal of Research of the National Bureau of Standards, **66A**, 179 (1962). The uncertainty of the assigned values of pH(S) is estimated not to exceed ± 0.005 unit from 0 to 50 °C. The values listed below apply only to the lots here certified. Minor variation of pH(S) (of the order of a few thousandths of a unit) may be expected to occur between different lots.

The solution 0.025 molal with respect to both KH_2PO_4 and Na_2HPO_4 is recommended for the calibration of pH equipment. The pH(S) of this solution as a function of temperature is given below:

°C	pH(S)	°C	pH(S)	°C	pH(S)
0	6.982	20	6.878	37	6.839
5	6.949	25	6.863	40	6.836
10	6.921	30	6.851	45	6.832
15	6.898	35	6.842	50	6.831

For pH measurements in the physiologically important range pH 7 to 8 a solution 0.008695 molal in KH_2PO_4 and 0.03043 molal in Na_2HPO_4 is also useful. The pH(S) values for this solution as a function of a temperature follow: [See Journal of Research of the National Bureau of Standards, **65A**, 267 (1961).]

°C	pH(S)	°C	pH(S)	°C	pH(S)
0	7.534	20	7.430	37	7.392
5	7.501	25	7.415	40	7.388
10	7.472	30	7.403	45	7.385
15	7.449	35	7.394	50	7.384

The potassium dihydrogen phosphate was obtained from the Mallinkrodt Chemical Works of St. Louis, Missouri; the disodium hydrogen phosphate from the J. T. Baker Chemical Co., of Phillipsburg, New Jersey. The experimental work leading to the certification of this material was carried out by M. Paabo and B. R. Staples.

The overall direction and coordination of technical measurements leading to the certification were performed under the chairmanship of R. A. Durst.

The technical and support aspects involved in the preparation, certification, and issuance of this Standard Reference Material were coordinated through the Office of Standard Reference Materials by T. W. Mears.

Washington, D. C. 20234
September 1, 1970

J. Paul Cali, Acting Chief
Office of Standard Reference Materials

PREPARATION OF THE 0.025-MOLAL SOLUTION

Transfer 3.387 g of the potassium dihydrogen phosphate (186-I-c) and 3.533 g of the disodium hydrogen phosphate (186-II-c) to a 1-liter volumetric flask. Dissolve and fill to the mark with distilled water at 25 °C. The distilled water should not contain dissolved carbon dioxide and should have a conductivity no greater than 2×10^{-6} ohm^{-1} cm^{-1}. Carbon dioxide-free water can be prepared by boiling a good grade of distilled water for 10 minutes and guarding it with a soda-lime tube while cooling. The salts should be dried for 2 hours at 110 °C to 130 °C before use. Although elaborate precautions to prevent contamination of the buffer solution with atmospheric carbon dioxide are usually unnecessary, the container should be kept tightly stoppered at all times when a sample is not actually being removed. (The figures given are weights in air.)

The solution should be replaced after a few weeks or sooner if molds or sediment appear, or if it has been exposed repeatedly to air containing carbon dioxide.

PREPARATION OF THE PHYSIOLOGICAL STANDARD SOLUTION

The physiological standard solution pH(S) = 7.415 at 25 °C, is prepared by transferring 1.179 g of potassium dihydrogen phosphate (186-I-c) and 4.303 g of disodium hydrogen phosphate (186-II-c) to a 1-liter volumetric flask, dissolving, and filling to the mark with water of the quality described above. (The figures given are weights in air.)

This buffer solution is more sensitive to contamination with carbon dioxide than is the 0.025-molal solution. If the solution is to maintain the assigned pH(S) for a few weeks, exclusion of carbon dioxide may be essential.

VIII. Education Committee*

and the
International Union of Pure and Applied Chemistry
Clinical Chemistry Division
Commission on Teaching of Clinical Chemistry**

* D. H. Curnow, Nedlands (AUS), Chairman 1976–81
C. G. Fraser, Bedford Park (AUS)
C. J. Porter, Toronto (CAN), Secretary 1976–81
M. K. Schwartz, New York, NY (USA), Chairman from 1982
H. G. J. Worth, Sutton in Ashfield (GBR), Secretary from 1982
O. Zinder, Haifa, (ISR)

** Titular Members

A. Defalque, Bruxelles (BEL)
M. Roth, Geneva (CHE)
M. K. Schwartz, New York, NY (USA)
H. G. J. Worth, Sutton in Ashfield (GBR)
O. Zinder, Haifa (ISR)

Associate Members

M. M. Abd El Kader, Cairo (EGY)
J. A. Correa, Rio de Janeiro (BRA)
B. Dastugue, (FRA)
D. Dotchev, Sofia (BGR)
A. L. Latner, Newcastle upon Tyne (GBR)
P. Lous, Copenhagen (DNK)
V. V. Menshikov, Moscow (SUN)
B. Strauss, Zagreb (YUG)

National Representatives

B. Christopherson, Oslo (NOR)
T. Kanno, (JPN)

A Scheme for a Two Year Postgraduate Course in Clinical Chemistry

Recommendations 1981[1])

Prepared for publication by C. J. Porter and D. H. Curnow

Preamble

The two year postgraduate course in clinical chemistry was developed by the IFCC Education Committee and the IUPAC Commission on Teaching of Clinical Chemistry, in response to a recognized need for clinical chemistry supervisors capable of undertaking responsible roles in clinical chemistry laboratories under the direction of the laboratory director. The need for a well trained "middle level" clinical chemist has become apparent in certain countries. The recommended course provides guidelines of a purposefully flexible nature so that they may be adapted to the needs of a particular country.

The course places primary emphasis on the technical/analytical and managerial aspects of clinical chemistry with sufficient education and training in clinical biochemistry to enable the candidate to gain a basic understanding of the interpretive aspects of clinical chemistry. It is not intended that the course brings the student to the level of competence required of the more highly qualified clinical chemist capable of overall direction of a clinical chemistry laboratory. Programs for training laboratory directors are described elsewhere (1,2) and generally require more in-depth education, training and experience over a longer period of time.

It would be expected that the establishment of the two year postgraduate course would be followed by the development of more advanced programs leading to the higher qualifications required of a laboratory director.

References

1. Rubin, M. & Lous, P. (1977) Education and Training for Clinical Chemistry, M. T. P. Press, Lancaster, England.
2. Post-graduate Training in Clinical Chemistry, 2nd Ed. 1979, prepared by the Association of Clinical Biochemists' Education Committee, U. K.

[1]) Approved by the Council of the International Federation of Clinical Chemistry 1982–11–01.
A provisional document was published in 1980 in Clin. Chim. Acta *103*, F239–F247; J. Clin. Chem. Clin. Biochem. *18*, 439–444 and Clin. Chem. *26*, 1746–1753.

Contents

1. Preface

The clinical chemist

That the Clinical Chemist plays an essential role in the diagnosis and management of patients is now widely accepted. The IFCC Committee on Education and the IUPAC Commission on Teaching of Clinical Chemistry considered that it would be helpful to prepare a syllabus for a training program, suitable for those who supervise service laboratories. The course outline is for graduates in an appropriate science subject.

The Clinical Chemist fulfills many roles. Perhaps first and foremost he must be a reliable and respected analyst providing his results with the speed demanded by the condition of the patient and the suspected diagnosis. With this role in mind much of the suggested course is concerned with analytical techniques. The Clinical Chemist is, however, in the vanguard of the scientists who play an increasingly important role in the multidisciplinary team which is involved in diagnosis and management in modern medicine. It would, therefore, be more than a pity if the Clinical Chemist holding a responsible position in the service laboratory were merely to be considered by his medical colleagues as a reliable analyst. Equally important is an understanding of the implications of his analytical findings and an ability to convey to his medical colleagues an integrated interpretation of clinical and analytical aspects.

The role of the Clinical Chemist will vary considerably according to the circumstances in which he finds himself. He may be employed in a large modern laboratory in an industrial country or is perhaps stationed in a small community hospital in a developing country. In any event he should be regarded by his medical colleagues as a person with a lively mind who is ready to give his opinion concerning the interpretation of laboratory results for which he is responsible. He should remember that he will probably have a much better concept of the quantitative significance of his results than will his clinical colleagues and for this reason, he must realize that he should be his own severest critic.

For the Clinical Chemist to be prepared to offer suggestions concerning the interpretation of his results, he must have a sound background of medical biochemistry. The training program should provide an opportunity to remedy any shortcomings in this respect; e. g., selected reading and attendance at ward rounds, case discussions, clinico-pathology conferences and special meetings.

Since Clinical Chemistry is a rapidly growing subject no course can be an end in itself – rather it should be regarded as a beginning which lays the background and encourages the student to develop the habit of self-instruction. We would like to suggest, therefore, that the student devises a reading program of the journals appropriate to his continuing education. To keep abreast of current literature, fertilizing the mind with new ideas, is essential if the Clinical Chemist is to gain and maintain the respect of his clinical colleagues.

2. Course outline

A two year program leading to an M.Sc or Diploma in Clinical Chemistry is proposed as part of the training of clinical chemistry specialists capable of undertaking supervisory roles in clinical chemistry service laboratories. The course places primary emphasis on the development of technical-analytical and managerial skills with adequate but relatively less emphasis on the clinical-interpretative aspects of clinical biochemistry. Graduates of such a program would be expected to have advanced capabilities in such areas as method development and evaluation, quality control, trouble shooting and minor repair of general laboratory equipment and general organization of workloads and technical staff. These duties would be carried out under the immediate direction of the laboratory director or qualified alternate.

It is recommended that the primary faculty and facilities be located centrally in a university department which for preference has service responsibilities, such as a department of Clinical Biochemistry in a teaching hospital. However, other settings which may be suitable include departments of Chemistry, Biochemistry, or Pharmacy provided that there are close links with a general hospital. Lectures and tutorials would be organized and presented by the university department with definitive interaction with the facilities of university departments of Chemistry, Biochemistry, Biology, Medicine and the clinical chemistry laboratories in affiliated hospitals. An advisory council composed of university faculty and hospital staff should be established to take responsibility for the development, coordination and conduct of the program. Additionally, it is expected that the IFCC through its Committee on Education in Clinical Chemistry, if requested, might assist in the development and evaluation of the program and in arranging for lectures and seminars by outstanding visiting lecturers.

It should be emphasized that the course to be described will require a full-time, two year registration and follows a structured curriculum, as distinct from the more general program of training for the Clinical Chemist which must last longer that two years and encompass additional aspects, e. g., hospital organization and administration, functional interrelationships of the medical and laboratory disciplines, etc.

Course objectives

1. To teach the fundamental principles of clinical chemistry through a systematic course of lectures, and tutorials.
2. To ensure that candidates achieve a high degree of practical competence in routine analyses and techniques through a program of supervised training in a hospital clinical chemistry laboratory.
3. To develop special skills in the use, maintenance, trouble-shooting, and minor repair of analytical instrumentation and general laboratory equipment.
4. To train students in the techniques of method development, method evaluation, quality analyses and control, laboratory management and assessment of medical usefulness of tests.
5. To provide a fundamental understanding of the clinical aspects of work performed in the clinical chemistry laboratory, through lectures, selected reading and attendance at appropriate ward rounds.
6. To stimulate the student to explore, by means of basic, applied or clinical research, certain aspects of the biochemical changes that accompany disease processes.

Prerequisites

Candidates must have a university degree or equivalent qualification in chemistry, biochemistry, medicine, pharmacy, or medical technology. Undergraduate work will usually have included courses in mathematics (including basic statistics), physics, biology, general, organic, analytical and physical chemistry, biochemistry and physiology.

Structure of course

Student(s) would be assigned to a hospital clinical chemistry laboratory where the director of the laboratory would act as supervisor of the student(s) working in his laboratory. A structured course of lectures and demonstrations should run through the whole period. If the services of foreign teachers are called upon, it would be more practical to present the course in blocks of one or two weeks. In any case, the courses should be integrated as much as possible with the in-service training program.

Lecture course

An outline of the program is provided in the Syllabus (below). Much of this material can be covered informally during the in-service part of the program in the form of seminars and discussion periods with the director of the program and/or senior clinical chemists. A formal lecture program, however, should be developed and presented centrally to all students in the program. A lecture course which integrates analytical chemistry and biochemistry with physiology and "pathological chemistry" should be based on the topics outlined in Section 3.6 of the Syllabus. This course should provide an appreciation of the biochemical and physiological factors involved in the maintenance and alteration of organ and tissue function. Equally important, the student should gain an understanding of the significance and limitations of the laboratory determinations involved. Supplementary lectures or seminars by outstanding visitors should be encouraged.

It is recognized that the particular needs for clinical chemistry services and the resources available differ from country to country. For these reasons the proposed course outline (Syllabus) should be regarded as a guide for the development of a program appropriate to a particular country's needs and resources.

In-service training

Students would systematically rotate through each section of the routine service laboratory to gain experience in the practical aspects of clinical chemistry. The student's progress should be evaluated weekly through informal discussion with the laboratory director and other members of the professional staff as available. A formal evaluation of the student's progress should be conducted at six monthly intervals.

When the student has demonstrated that he has acquired the required level of practical competence and knowledge, he should be given short term supervisory responsibility for the work of that area or section before moving on to his next assignment. It would be desirable that he undertake some degree of responsibility for work performed after regular working hours, e. g., emergency services.

During the second year of in-service training, the student would undertake projects in method development, method evaluation, quality assurance and control, and laboratory management, in addition to acting in a back-up supervisory capacity in those areas for which he had demonstrated a high degree of competence, including instrument maintenance, trouble-shooting and repair. The projects assigned should be of relatively short duration in order to provide a broad experience, and ensure that the student develops many of the skills required of a laboratory supervisor. During this second year the student should improve his teaching skills by involvement in staff and student training and continuing education programs.

Where certain techniques are not carried out in a laboratory, arrangements should be made for the student to gain experience in those particular techniques in another laboratory.

The in-service training program should specify regular involvement in certain clinical activities, such as ward rounds, case discussions and clinico-pathology conferences, to enable the post-graduate student to appreciate more fully the advantage of being an active member of the multidisciplinary clinical-care team. Another activity that will develop skills in interpreting laboratory data includes involvement in "reporting" laboratory tests. Where appropriate, additional clinical information should be sought to enable the laboratory results to be correctly interpreted in relation to the specific clinical setting.

Reading list

Students are expected to do both general background reading and undertake in-depth study of specific topics. The student should learn to look at a scientific article in a critical way and develop an ability to assess the validity of the results reported and their possible application to his own work. A list of books, journals and special articles currently in use in different countries has been compiled and is available on request[2]).

[2]) Dr. C. J. Porter, Department of Clinical Biochemistry, Toronto General Biochemistry, Toronto General Hospital, 200 Elizabeth Street, Toronto, Ontario M5G 117 Canada.

Examination requirements

Students will be required to pass written, oral and practical examinations and present satisfactory progress reports on their in-service training and project work in order to obtain the M.Sc or post-graduate Diploma in Clinical Chemistry. Students should be encouraged to take certifying examinations in those countries where they are established. These usually require one or two years of practical experience in an approved clinical chemistry laboratory.

3. Syllabus

The syllabus provides a summary of the material considered to be essential for a two year post-graduate course in Clinical Chemistry. This should be used as a flexible guide for the development of a more detailed curriculum approporiate to the needs and resources available to a particular country. As it is recognized that in some countries responsibility for hematology and/or microbiology or toxicology is undertaken by the clinical chemistry service, it would be expected that the relative emphasis on these subjects will vary according to local practice. The development and implementation of a training program based on the proposed guidelines should be undertaken by a local or national faculty.

The curriculum should be developed around a didactic program consisting of lectures, tutorials and informal practical teaching sessions in the laboratory. Some of the topics might be dealt with by visiting lecturers or visits to other training centres.

The subject material for the proposed post-graduate course is organized under the following headings:

3.1 General laboratory procedures
3.2 Techniques and principles used in clinical chemistry
3.3 Instrumentation
3.4 Analyses
3.5 Special projects
3.6 Clinical biochemistry
3.7 Clinical medicine
3.8 Laboratory organization and management

3.1 General laboratory procedure

The following procedures are considered to be essential to the operation of a routine service clinical chemistry laboratory. Most of these procedures will be learned through practical experience during in-service training supplemented with informal teaching, demonstrations and selected reading.

Specimen collection, preservation and preparation for analysis; constituent stability; documentation and sample flow systems.

Laboratory statistics: concepts of reference values and biological variation; confidence limits, t-test, F-test, χ^2-test, regression analysis, non-parametric tests, precision, accuracy.

Chemical arithmetic, e. g., molar solutions, molar absorption, pH calculations, radioactive decay, isotope dilution.

Internal quality control, external quality assurance, quality assurance, reference methods, control of non-analytical aspects of laboratory procedures.

Quantities and units: conventional and SI units – conversion.

Laboratory requisitions and reporting systems including data processing and use of computers laboratory records, workload statistics – legal aspects.

Reagents and apparatus: sources of supply, cost analysis, techniques for assessing quality of reagents and kits.

Materials for control of precision and accuracy: sources of supply, techniques for assigning appropriate values.

3.2 Techniques and principles used in clinical chemistry

It would be expected that most candidates would have some undergraduate exposure to many of the topics listed here. Instruction should be provided to ensure theoretical and if necessary, practical knowledge.

General techniques: Preparation, purification and storage of reagents and solvents; solvent extraction, selection of buffers, dialysis, filtration and ultra-filtration, concentration; preparation of derivatives, preparation of protein-free filtrates.

Chromatography: Various supporting media, e. g., paper membranes, gels, thin layer, ion exchange resins; gas liquid chromatography and high performance liquid chromatography.

Electrophoresis: Conventional one and two dimensional electrophoresis: various supporting media.

Photometry, spectrophotometry and fluorometry: Techniques of analysis by visible, ultraviolet, infrared and fluorometric procedures. Applications in clinical chemistry.

Enzymology: Methods of estimation of enzymes and isoenzymes (fixed time incubation and continuous methods). Units, standardization, stability.

Immunochemical assay: Immunodiffusion, immunoelectrophoresis, competitive binding techniques, radioimmunoassay, EMIT, ELISA and receptor assays.

Radioisotopes: General principles; counting techniques; safety aspects and legal requirements for use and disposal. Applications and procedures in diagnostic tests (radioimmunoassay).

Microchemical analysis: Experience with small sample and reagent volumes (pediatric clinical chemistry).

3.3 Instrumentation

The following instruments and apparatus are considered to be essential for operation of a routine clinical chemistry laboratory. Candidates should understand the theoretical and practical aspects, be able to set up and operate the equipment, trouble-shoot and perform minor repairs. They should be able to establish the correctness of calibration of the instruments and have knowledge of their applications and limitations.

General laboratory equipment such as centrifuges, water baths, balances, microscopes, pipetting devices, automatic dispensers.
Water purification systems – stills, de-ionizers, methods of checking and monitoring water quality.
Spectrometric instruments – absorption, flame emission, atomic absorption spectrometers and fluorometers.
pH meters
Blood gas apparatus
Electrophoresis and chromatographic equipment
Gas and high performance liquid chromatographic equipment
Continuous flow analyzers and discrete sample analyzers
Electrometric titrators
Kinetic analyzers

The instruments and apparatus listed are in common use, but not necessarily found in all laboratories. To gain experience with instruments not available in the assigned training laboratory, arrangements should be made for the student to rotate to a hospital laboratory where such instruments are routinely used.

3.4 Analyses

Listed below in alphabetic order are some of the components commonly examined in an appropriate biological fluid (e. g., serum, plasma, cerebrospinal fluid, etc.) in a routine service laboratory. The student should be familiar with the theoretical principles of various methods used for measurement of these components and be competent in the performance of the particular method employed in the laboratory to which he is assigned. The student should also be familiar with the procedures employed for selection and evaluation of methods, not only for precision, accuracy, speed, simplicity, cost, etc., but also for sensitivity and specificity. This obviously relates to the selection of tests for specific clinical purposes and the assessment of the medical usefulness of tests.

Acid phosphatase
Alkaline phosphatase
α-Amylase
Alanine aminotransferase
Albumin
Aspartate aminotransferase
Barbiturate
Bilirubins
Bilirubin esters
Calcium
Carbamide
Carbonate + carbon dioxide
Carbon dioxide (pCO_2)
Catecholamines
Chloride
Cholesterol and esters
Coproporphyrin (I + III)
Cortisol
Creatininium
Digoxin
Dioxygen (pO_2)
Diphenylhydantoin
Oestrogens
γ-Glutamyl transferase
Glucose
Haemoglobin
Hydrogen carbonate ion
17-Hydroxycorticosteroids
Hydroxymethyl mandelate
Iron (II + III)
Lactate dehydrogenase
Lithium ion
Magnesium
Osmolality
17-Oxogenic steroids
17-Oxosteroids
pH
Porphobilinogen
Potassium ion
Protein
Salicylate
Sodium ion
Thyroid stimulating hormone
Thyroxine
Transferrin
Triglycerides
Triiodothyronine
Urate
Urobilinogens
Uroporphyrin (I + III)

In addition, many laboratories will carry out procedures for the analysis of urine and stools, and for the assessment of gastric secretion, serum proteins by electrophoresis and for the assessment of pregnancy.

3.5 Special projects

Short-term projects involving a considerable number of the skills required of a laboratory supervisor (analytical, instrumental, evaluative, managerial, organizational) should be undertaken to encourage initiative and independence. The student should acquire an ability for clear report writing.

3.6 Clinical biochemistry

Candidates should acquire a general knowledge of the interpretative aspects of clinical biochemistry. A central course of lectures should be developed which integrate analytical chemistry and biochemistry with physiology and physiological chemistry. Case histories selected to illustrate biochemical dearrangements in disease states should be integrated with the main course to give the student an opportunity to develop skills in the interpretation of laboratory results. These sessions should also relate to the selection and evaluation tests and their medical usefulness.

It is suggested that the course in clinical biochemistry should include a general treatment of the topics listed below. The diagnostic tests listed in section 3.4 provide a general indication of the extent of coverage of the topics selected. More detailed guidelines are available from the Secretary of the Commission/Committee, on request.

Water and ionic balance
Renal function
Acid-base metabolism
Carbohydrate metabolism
Plasma proteins
Lipids and lipoproteins
Gastro-intestinal function
Liver function
Calcium, phosphorus and magnesium metabolism
Endocrine function
Common hereditary disorders
Nutrition
Purine metabolism
Clinical chemistry of pregnancy
Haeme and porphyrins
*Haematology and coagulation
Clinical enzymology
Clinical pharmacology and pharmacokinetics
Toxicology

* The student should acquire a basic understanding of the general principles of haematology and blood coagulation. In countries where haematology is a responsibility of the clinical chemist, however, considerably greater emphasis on this component of the course will be required.

3.7 Clinical medicine

The science graduate entering clinical chemistry should seek every opportunity to familiarize himself with the relevant aspects of clinical medicine in order to communicate effectively with the medical staff. Knowledge about clinical medicine is also important for the medical graduate coming directly from the medical school. The supervisory staff should include a medically qualified clinical chemist if available. If not, satisfactory alternative arrangements will need to be made.

3.8 Laboratory organization and management

Laboratory safety:	Mechanical, electrical, chemical, biological, fire and radiation hazards.
Laboratory work flow:	Planning for efficiency; work flow and communication systems.
Budget:	Preparation and monitoring.
Workload:	Data processing (manual and/or computer).
Quality assurance:	Implementation, monitoring and performance evaluation.
Technical staff:	Education, training, performance assessment, work assignment.

Acknowledgements

The contribution of National Representatives and other individuals is gratefully acknowledged, especially the following: F. A. Tares Taie (Argentina); P. M. Dennis, C. Fraser, D. W. Thomas (Australia); L. A. Wright, D. Mickle, L. Allen (Canada); M. Monair (Chile); J. Homolka (Czechoslowakia); C. Davila Saa (Ecuador); O. Siggaard Andersen, F. Damgaard-Pedersen, R. Dybkaer (Denmark); R. Vihko (Finland); D. W. Scheuch (GDR); B. Ringelhann (Hungary); W. C. Love (Ireland); Y. Yamamura (Japan); M. L. Castillo de Sanchez (Mexico); W. N. Chisnall, J. Mattingly (New Zealand); A. P. Jansen (Netherlands); O. Bormer (Norway); J. Prokopowica (Poland); K. W. Arntsen (South Africa); A. Kallner (Sweden); P. Broughton, S. S. Brown, P. N. Campbell, A. L. Miller (UK); V. S. Jather, A. Kaplan, I. Sunshine (U.S.A.); I. Ruzdic (Yugoslavia).

Bibliography of Published IFCC Documents

Grouped According to Committee, Expert Panel etc.

Regulations for Preparation

Current General Regulations for Preparation of IFCC Recommendations (1975–07–10)
Rubin, M., Büttner, J. & Saris, N.-E.
J. Clin. Chem. Clin. Biochem. *15* (1977) 52–55

Committee on Education

IUPAC Commission on Teaching of Clinical Chemistry
Education and Training for Clinical Chemistry
(Eds. M. Rubin & P. Lous)
M.T.P. Press, Lancaster (1977)

The Education and Training of Personnel for Clinical Chemistry
Rubin, M. & Lous, P.
Clin. Chim. Acta *61* (1975) F1–F10, *following p. 112*
J. Clin. Chem. Clin. Biochem. *13* (1975) 465–470

Provisional Recommendation (1979) on a Scheme for a Two Year Postgraduate Course in Clinical Chemistry
Porter, C. J. & Curnow, D. H.
Clin. Chim. Acta *103* (1980) 239F–147F
J. Clin. Chem. Clin. Biochem. *18* (1980) 439–444
Clin. Chem. *26* (1980) 1748–1753

A Scheme for a Two Year Postgraduate Course in Clinical Chemistry
Porter, C. J. & Curnow, D. H.
J. Clin. Chem. Clin. Biochem. *21* (1983) 185–191
Clin. Chim. Acta *131* (1983) 351F–359F
Pure & Appl. Chem. *55* (1983) 557–564

Expert Group on Diagnostic Kits and Reagents

Provisional Recommendation (1978) on Evaluation of Diagnostic Kits
Part 1. Recommendations for Specifications on Labelling of Clinical Laboratory Materials
Rubin, M., Barnett, R. N., Bayse, D., Beutler, E., Brown, S. S., Logan, J. E., Reimer, C. B., Westgard, J. O. & Wilding, P.
Clin. Chim. Acta *95* (1979) 155F–162F
J. Clin. Chem. Clin. Biochem. *17* (1979) 657–661
Clin. Chem. *25* (1979) 1499–1502

Provisional Recommendation (1978) on Evaluation of Diagnostic Kits
Part 2. Guidelines for the Evaluations of Clinical Chemistry Kits
Rubin, M., Barnett, R. N., Bayse, D., Beutler, E., Brown, S. S., Logan, J. E., Reimer, C. B., Westgard, J. O. & Wilding, P.
Clin. Chim. Acta *95* (1979) 163F–168F
J. Clin. Chem. Clin. Biochem. *17* (1979) 661–664
Clin. Chem. *25* (1979) 1503–1505

Revised Recommendation (1983) on Evaluation of Diagnostic Kits
Part 1. Recommendations for Specifications on Labelling of Clinical Laboratory Materials
Logan, J. E.
J. Clin. Chem. Clin. Biochem. *21* (1983) 893–898
Clin. Chim. Acta *137* (1984) 371F–379F

Revised Recommendation (1983) on Evaluation of Diagnostic Kits Part 2. Guidelines for the Evaluation of Clinical Chemistry Kits Logan, J. E.	J. Clin. Chem. Clin. Biochem. *21* (1983) 899–902 Clin. Chim. Acta *137* (1984) 381F–386F

Expert Panel on Evaluation of Diagnostic Reagent Sets

The World Health Organization

IFCC/WHO Principles and Recommendations on Evaluation of Diagnostic Reagents Sets Used in Health Laboratories with Limited Resources Part I. Selection and Evaluation Using Reference Materials. General Conditions Part II. Evaluation of Performance Using Reference Materials of Analytes Commonly Determined in Blood Serum of Plasma Logan, J. E., Bayse, D. D., Koedam, J. C., Mather, A. & Wilding, P.	LAB/83.1.
Appendix A: Glucose	LAB/83.2.
Appendix B: Urea	LAB/83.3.
Appendix C: Total Bilirubin	LAB/83.4.
Appendix D: Total Protein	LAB/83.5.
Appendix E: Albumin	LAB/83.6.

Expert Panel on Enzymes

Provisional Recommendation (1974) on IFCC Methods for the Measurement of Catalytic Concentration of Enzymes Bowers, Jr., G. N., Bergmeyer, H. U. & Moss, D. W.	Clin. Chim. Acta *61* (1975) F11–F24, *following p. 238* J. Clin. Chem. Clin. Biochem. *13* (1975) 471–478 Clin. Chem. *22* (1976) 384–391
IFCC Methods for the Measurement of Catalytic Concentration of Enzymes Part 1. General Considerations Concerning the Determination of Catalytic Concentration of an Enzyme in Blood Serum or Plasma of Man Bowers, Jr., G. N., Bergmeyer, H. U., Hørder, M. & Moss, D. W.	Clin. Chim. Acta *98* (1979) 163F–174F J. Clin. Chem. Clin. Biochem. *18* (1980) 89–95
Provisional Recommendations on IFCC Methods for the Measurement of Catalytic Concentrations of Enzymes Part 2. IFCC Method for Aspartate Aminotransferase Bergmeyer, H. U., Bowers, Jr., G. N., Hørder, M. & Moss, D. W.	Clin. Chim. Acta *70* (1976) F19–F42, *following p. 336* J. Clin. Chem. Clin. Biochem. *15* (1977) 39–51 Clin. Chem. *23* (1977) 887–899
Provisional Recommendations on IFCC Methods for the Measurement of Catalytic Concentrations of Enzymes Part 3. Revised IFCC Method for Aspartate Aminotransferase (1977) Bergmeyer, H. U., Hørder, M. & Moss, D. W.	Clin. Chim. Acta *80* (1977) F21–F22, *following p. 394* J. Clin. Chem. Clin. Biochem. *15* (1977) 719–720 Clin. Chem. *24* (1978) 720–721
IFCC Methods for the Measurements of Catalytic Concentration of Enzymes Part 3*. IFCC Method Alanine Aminotransferase Bergmeyer, H. U. & Hørder, M.	J. Clin. Chem. Clin. Biochem. *18* (1980) 521–534 Clin. Chim. Acta *105* (1980) 147F–172F
Report on the Survey of Opinions Concerning Preferred Incubation Temperatures for Measurement of Enzymes (and Possibly Other Components) in Clinical Chemistry Haeckel, R., Hørder, M. & Zender, R.	J. Clin. Chem. Clin. Biochem. *20* (1982) 947–958 Clin. Chim. Acta *129* (1983) 147F–249F

*) Subsequently defined as "Addendum to Part 2".

IFCC Methods for the Measurement of Catalytic Concentration of Enzymes Part 4. IFCC Method for γ-Glutamyltransferase Shaw, L. M., Strømme, J. H., London, J. L. & Theodorsen, L.	J. Clin. Chem. Clin. Biochem. *21* (1983) 633–646 Clin. Chim. Acta *135* (1983) 315F–338F
IFCC Methods for the Measurement of Catalytic Concentration of Enzymes Part 5. IFCC Method for Alkaline Phosphatase Tietz, N. W., Rinker, A. D. & Shaw, L. M.	J. Clin. Chem. Clin. Biochem. *21* (1983) 731–748 Clin. Chim. Acta *135* (1983) 339F–367F

Expert Panel on Instrumentation

Provisional Guidelines (1979) for Listing Specifications of Spectrometers in Clinical Chemistry Haeckel, R., Collombel, Ch., Geary, T. D., Mitchell, F. L., Nadeau, R. G. & Okuda, K.	J. Clin. Chem. Clin. Biochem. *18* (1980) 445–449 Clin. Chim. Acta *103* (1980) 249F–258F Clin. Chem. *27* (1981) 187–191
Provisional Guidelines (1980) for Listing Specifications of Clinical Chemistry Analysers Okuda, K.	J. Clin. Chem. Clin. Biochem. *18* (1980) 947–951 Clin. Biochem. *13* (1980) 244–248 Clin. Chim. Acta *119* (1982) 351F–362F
Provisional Guidelines (1981) for Listing Specifications of Flame Emission Spectrometers Bechtler, G., Epstein, M. S., Geary, T. D., Havemann, W. & Attoe, P.	J. Clin. Chem. Clin. Biochem. *20* (1982) 256–261 Clin. Chim. Acta *122* (1982) 111F–115F
Provisional Guidelines (1981) for Listing Specifications of Atomic Absorption Spectrometers Epstein, M. S., Geary, T. D., Gower, G., Tausch, W., Mills, K. J. & Polt, D.	J. Clin. Chem. Clin. Biochem. *20* (1982) 263–266 Clin. Chim. Acta *122* (1982) 117F–123F Clin. Chem. Newsletter *2* (1982) 67–73
Bibliography on Evaluation of Instrumentation in Clinical Chemistry Donohoe, G. A., Geary, T. D. & Jennings, R. D.	J. Clin. Chem. Clin. Biochem. *20* (1982) 931–945 Clin. Chim. Acta *127* (1983) 425F–439F Clin. Chim. Newsletter *3* (1983) 3–13
Report on the Survey of Opinions concerning Preferred Incubation Temperatures for Measurement of Enzymes (and Possibly Other Components) in Clinical Chemistry Haeckel, R., Hørder, M. & Zender, R.	J. Clin. Chem. Clin. Biochem. *20* (1982) 947–958 Clin. Chim. Acta *129* (1983) 247F–249F

Expert Panel on Nomenclature and Principles of Quality Control in Clinical Chemistry

Provisional Recommendation on Quality Control in Clinical Chemistry Part 1. General Principles and Terminology Büttner, J., Borth, R., Boutwell, J. H. & Broughton, P. M. G.	Clin. Chim. Acta *63* (1975) F25–F38, *following p. 108* J. Clin. Chem. Clin. Biochem. *13* (1975) 523–531 Clin. Chem. *22* (1976) 532–540
Approved Recommendation (1978) on Quality Control in Clinical Chemistry Part 1. General Principles and Terminology Büttner, J., Borth, R., Boutwell, J. H., Broughton, P. M. G. & Bowyer, R. C.	J. Clin. Chem. Clin. Biochem. *18* (1980) 69–77 Clin. Chim. Acta *98* (1979) 129F–143F
Provisional Recommendation on Quality Control in Clinical Chemistry Part 2. Assessment of Analytical Methods for Routine Use Büttner, J., Borth, R., Boutwell, J. H., Broughton, P. M. G. & Bowyer, R. C.	Clin. Chim. Acta *69* (1976) F1–F17, *following p. 152* J. Clin. Chem. Clin. Biochem. *14* (1976) 265–275

Approved Recommendation (1978) on
Quality Control in Clinical Chemistry
Part 2. Assessment of Analytical Methods for Routine Use
Büttner, J., Borth, R., Boutwell, J. H.,
Broughton, P. M. G. & Bowyer, R. C.

Clin. Chim. Acta *98* (1979) 145F–162F
J. Clin. Chem. Clin. Biochem. *18* (1980) 78–88

Provisional Recommendation on
Quality Control in Clinical Chemistry
Part 3. Calibration and Control Materials
Büttner, J., Borth, R., Boutwell, J. H.,
Broughton, P. M. G. & Bowyer, R. C.

Clin. Chim. Acta *75* (1977) F11–F20, *following p. 180*
J. Clin. Chem. Clin. Biochem. *15* (1977) 233–238
Clin. Chem. *23* (1977) 1784–1789

Approved Recommendation (1979) on
Quality Control in Clinical Chemistry
Part 3. Calibration and Control Materials
Büttner, J., Borth, R., Boutwell, J. H.,
Broughton, P. M. G. & Bowyer, R. C.

J. Clin. Chem. Clin. Biochem. *18* (1980) 855–860
Clin. Chim. Acta *109* (1981) 105F–114F

Quality Control in Clinical Chemistry
Part 4. Internal Quality Control
Büttner, J., Borth, R., Boutwell, J. H.,
Broughton, P. M. G. & Bowyer, R. C.

J. Clin. Chem. Clin. Biochem. *18* (1980) 535–541
Clin. Chim. Acta *106* (1980) 109F–120F

Approved Recommendation (1983) on
Quality Control in Clinical Chemistry
Part 4. Internal Quality Control
Büttner, J., Borth, R., Boutwell, J. H.,
Broughton, P. M. G. & Bowyer, R. C.

J. Clin. Chem. Clin. Biochem. *21* (1983) 877–884

Provisional Recommendation on
Quality Control in Clinical Chemistry
Part 5. External Quality Control
Büttner, J., Borth, R., Boutwell, J. H.
Broughton, P. M. G. & Bowyer, R. C.

Clin. Chim. Acta *83* (1978) 189F–202F
J. Clin. Chem. Clin. Biochem. *16* (1978) 259–266
Clin. Chem. *24* (1978) 1213–1220

Approved Recommendation (1983) on
Quality Control in Clinical Chemistry
Part 5. External Quality Control
Büttner, J., Borth, R., Boutwell, J. H.,
Broughton, P. M. G. & Bowyer, R. C.

J. Clin. Chem. Clin. Biochem. *21* (1983) 885–892

Provisional Recommendation on
Quality Control in Clinical Chemistry
Part 6. Quality Requirements from
the Point of View of Health Care
Büttner, J., Borth, R., Boutwell, J. H.,
Broughton, P. M. G. & Bowyer, R. C.

Clin. Chim. Acta *74* (1977) F1–F9, *following p. 190*
J. Clin. Chem. Clin. Biochem. *15* (1977) 95–100
Clin. Chem. *23* (1977) 1066–1071

Approved Recommendation (1979) on
Quality Control in Clinical Chemistry
Part 6. Quality Requirements from
the Point of View of Health Care
Büttner, J., Borth, R., Boutwell, J. H.,
Broughton, P. M. G. & Bowyer, R. C.

J. Clin. Chem. Clin. Biochem. *18* (1980) 861–866
Clin. Chim. Acta *109* (1981) 115F–124F

Expert Panel on pH and Blood Gases

Physico-Chemical Quantities and Units
in Clinical Chemistry with Special Emphasis
on Activities and Activity Coefficients
Siggaard-Andersen, O., Durst, R. A. & Maas, A. H. J.

J. Clin. Chem. Clin. Biochem. *18* (1980) 829–854
Clin. Chim. Acta *108* (1980) 499F–539F

Reference Method of pH Measurement in Blood Maas, A. H. J., Weisberg, W. G., Zijlstra, W. G., Durst, R. A. & Siggaard-Andersen, O.	J. Clin. Chem. Clin. Biochem. *21* (1983) 313–321 Clin. Chim. Acta *134* (1983) 371F–386F J. Clin. Lab. Autom. *3* (1983) 263–268 Clin. Chem. NewsLett. *3* (1983) 48–53

Expert Panel on Proteins

Provisional Recommendation (1978) on Specification for Human Serum Albumin Standard Hobbs, J. R., Harboe, N. & Alper, C.	Clin. Chim. Acta *98* (1979) 175F–178F J. Clin. Chem. Clin. Biochem. *18* (1980) 96–98
The Human Serum IFCC 74/1 Hobbs, J. R., Harboe, M., Alper, C. & Johansson, B. G.	Clin. Chim. Acta *98* (1979) 179F–186F J. Clin. Chem. Clin. Biochem. *18* (1980) 99–103

Expert Panel on Quantities and Units

IUPAC Commission on Quantities and Units in Clinical Chemistry

Quantities and Units in Clinical Chemistry (Recommendation 1973) R. Dybkær	Pure & Appl. Chem. *37* (1974) 517–546
List of Quantities in Clinical Chemistry (Recommendation 1973) R. Dybkær	Pure & Appl. Chem. *37* (1974) 547–572
Provisional Recommendation (1978) on Quantities and Units in Clinical Chemistry – Optical Spectroscopy – Part 1. Theoretical Outline and General Quantities Rigg, J. C. & Zender, R.	Clin. Chim. Acta *95* (1979) 169F–188F J. Clin. Chem. Clin. Biochem. *17* (1979) 665–675 IUPAC Inf. Bull. *3* (1978) 141–260
Approved Recommendation (1978) Quantities and Units in Clinical Chemistry Dybkær, R.	Clin. Chim. Acta *96* (1979) 155F–183F J. Clin. Chem. Clin. Biochem. *17* (1979) 807–821 Pure & Appl. Chem. *51* (1979) 2451–2479
Approved Recommendation (1978) List of Quantities in Clinical Chemistry Dybkær, R.	Clin. Chim. Acta *96* (1979) 185F–204F J. Clin. Chem. Clin. Biochem. *17* (1979) 822–835 Pure & Appl. Chem. *51* (1979) 2481–2509
Physico-Chemical Quantities and Units in Clinical Chemistry with Special Emphasis on Activities and Activity Coefficients Siggaard-Andersen, O., Durst, R. A. & Maas, A. H. J.	J. Clin. Chem. Clin. Biochem. *18* (1980) 829–854 Clin. Chim. Acta *108* (1980) 499F–539F

Expert Panel on the Theory of Reference Values

Provisional Recommendation on the Theory of Reference Values (1978) Part 1. The Concept of Reference Values Gräsbeck, R., Siest, G., Wilding, P., Williams, G. Z. & Whitehead, T. P.	Clin. Chim. Acta *87* (1978) 459F–465F J. Clin. Chem. Clin. Biochem. *17* (1979) 337–339 Clin. Chem. *25* (1979) 1506–1508
Part 2. Selection of Individuals for the Production of Reference Values. PetitClerc, C. & Wilding, P.	J. Clin. Chem. Clin. Biochem. *22* (1984) 203–208
The Theory of Reference Values Part 6. Presentation of Observed Values Related to Reference Values Dybkær, R.	J. Clin. Chem. Clin. Biochem. *20* (1982) 841–845 Clin. Chim. Acta *127* (1983) 331F–448F Clin. Chem. Newsletter *3* (1982) 115–119

The Theory of Reference Values
Part 5. Statistical Treatment of Collected Reference Values, Determination of Reference Limits
Solberg, H. E.
J. Clin. Chem. Clin. Biochem. *21* (1983) 749–760
Clin. Chim. Acta *137* (1984) 95F–114F

Expert Panel on Drug Effects in Clinical Chemistry

Part 1. The Basic Concepts
Siest, G. & Dawkins, J.
J. Clin. Chem. Clin. Biochem. *22* (1984) 271–274

Part 2. Guidelines for Evaluation of Analytical Interference
Galteau, M. M. & Siest, G.
J. Clin. Chem. Clin. Biochem. *22* (1984) 275–279

List of Authors

G. Bechtler, Eppendorf Gerätebau GmbH, Barkhausenweg 1, D-2000 Hamburg 63, F.R. of Germany

H. U. Bergmeyer, Boehringer Mannheim GmbH, Biochemisches Werk Tutzing, Bahnhofstr. 5, D-8132 Tutzing, F. R. of Germany

D. H. Curnow, Department of Clinical Biochemistry, Queen Elizabeth II Medical Center, Medlands, W.A., Australia

R. A. Durst, Organic Analytical Research Division, Center for Analytical Chemistry, United States Department of Commerce, National Bureau of Standards, Washington, DC 20234, USA

R. Dybkær, Department of Clinical Chemistry, Frederiksberg Hospital, Nordre Fasanvej 59, DK-2000 Copenhagen, Denmark

R. Haeckel, Zentralkrankenhaus, St.-Jürgen-Straße, Institut für Labor. Medizin, D-2800 Bremen 1, F. R. of Germany

M. Hørder, Department of Clinical Chemistry and International Medicine B, Odense University Hospital, DK-5000 Odense C, Denmark

H. Lappe, Beckman Instruments GmbH, Frankfurter Ring 115, D-8000 München 40, F.R. of Germany

J. E. Logan, Laboratory Centre for Disease Control, Tunney's Pasture, Ottawa, Ontario K1A OL2, Canada

A. H. J. Maas, Clinical Chemistry Laboratory, Departments of Cardiac Surgery and Cardiology, University Hospital, Catharijnesingel 101, NL-3500 CG Utrecht, The Netherlands

K. Okuda, Central Clinical Laboratory, Osaka City University, Osaka, Japan

C. J. Porter, Department of Clinical Biochemistry, Toronto General Hospital, Toronto M5G 1L7, Canada

R. Rej, Director of Clinical Chemistry, State of New York, Department of Health, Center for Laboratories and Research, Albany N.Y. 12202, USA

O. Siggaard-Andersen, Department of Clinical Chemistry Københavns Amts Sygehus Herlev, Herlev Ringvej, DK-2730 Herlev, Denmark

W. Tausch †

H. F. Weisberg, The Ida Soref – David and Ruth Coleman Department of Pathology and Laboratory Medicine, Mount Sinai Medical Center, 950 North Twelfth Street, P.O. Box 342, Milwaukee, Wi 53201 USA

W. G. Zijlstra, Department of Physiology, University of Groningen, NL-9712 KZ Groningen, The Netherlands

Journal of Clinical Chemistry and Clinical Biochemistry

Zeitschrift für Klinische Chemie und Klinische Biochemie

Gemeinsames Organ der Deutschen, der Niederländischen, der Österreichischen und der Schweizerischen Gesellschaft für Klinische Chemie

This journal publishes all IFCC Recommendations regularly

Subscription Information
1985. Vol. 23 (12 issues): DM 666,–; approx. US $246,75. Single issue: DM 65,–; approx. US $24,25

Prices are subject to change without notice

W DE G Walter de Gruyter · Berlin · New York

Verlag Walter de Gruyter & Co., Genthiner Str. 13, D-1000 Berlin 30, Telefon: (0 30) 2 60 05-0
Walter de Gruyter, Inc., 200 Saw Mill River Road, Hawthorne, N. Y. 10532, Telephone: (914) 747-0110

FSC
www.fsc.org
MIX
Papier | Fördert gute Waldnutzung
FSC® C083411

Kolibri 360 GmbH
Ferdinand-Jühlke-Straße 7
99095 Erfurt, Deutschland
produktsicherheit@kolibri360.de